RAND NATIONAL DEFENSE RESEARCH INSTITUTE

Baselining Defense Acquisition

Philip S. Anton, Tim Conley, Irv Blickstein, Austin ⌐
William Shelton, Sarah Harting

Prepared for the Office of the Secretary of Defense

For more information on this publication, visit www.rand.org/t/RR2814

Library of Congress Cataloging-in-Publication Data is available for this publication.
ISBN: 978-1-9774-0202-8

Published by the RAND Corporation, Santa Monica, Calif.

© Copyright 2019 RAND Corporation

RAND® is a registered trademark.

Support RAND

Make a tax-deductible charitable contribution at
www.rand.org/giving/contribute

www.rand.org

Preface

In a February 2017 memorandum titled "Establishment of Cross-Functional Teams to Address Improved Mission Effectiveness and Efficiencies in the DoD," the Secretary of Defense stated his intent to field a larger, more capable, and more lethal Joint force. To fund this expansion, he directed the U.S. Department of Defense (DoD) to make its business operations more efficient by identifying business services and tasks that could be consolidated across the DoD. His direction focused specifically on human resource management, financial management, real-property management, acquisition and contract management, logistics and supply-chain management, health care management, base services, and cyber and information technology management.

In response, the then–Office of the Under Secretary of Defense (Acquisition, Technology and Logistics [OUSD(AT&L)])[1] asked the National Defense Research Institute (NDRI), a federally funded research and development center operated by the RAND Corporation, to identify changes to the DoD's processes for acquisition and procurement, real property management, and logistics and supply-chain management that might increase efficiency while preserving needed performance. As part of that effort, NDRI was asked to construct a baseline of government acquisition and procurement functions, including a functional decomposition and approaches for estimating the cost of executing the government acquisition enterprise. This report documents that baseline effort. Given the resource allocation implications, this baseline helps create clarity of assumptions in dealing with data and their sources.

The figures and conclusions in the Summary section of this report are intended for decisionmakers interested in how much the DoD spends executing the government functions in the acquisition enterprise. The detailed technical discussion of data sources and algorithms in the body of this report are intended more for analysts (and the few leaders) who have functional expertise and want to understand how we arrived at our results.

This research was sponsored by the Director of Acquisition, Resources, and Analysis in the then-OUSD(AT&L) and conducted within the Acquisition and Technology Policy Center of the RAND National Defense Research Institute, a federally funded research and development center sponsored by the Office of the Secretary of Defense, the Joint Staff, the Unified Combatant Commands, the Navy, the Marine Corps, the defense agencies, and the defense Intelligence Community.

---

[1] OUSD(AT&L) has since been replaced on February 1, 2018, with the OUSD(Research and Engineering) and the OUSD(Acquisition and Sustainment).

For more information on RAND's Acquisition and Technology Policy Center, see www.rand.org/nsrd/ndri/centers/atp or contact the director (contact information is provided on the webpage).

Contents

Figures and Tables

Figures

Tables

Summary

As part of the effort to identify efficiencies in the acquisition system, the Office of the Secretary of Defense (OSD) asked the National Defense Research Institute, a federally funded research and development center operated by the RAND Corporation, to explore ways for clarifying government execution functions and costs in the acquisition enterprise. This baseline entails listing and defining the basic government acquisition functions, exploring practical data sources for determining the cost of executing these government functions, and testing these data sources to obtain cost estimates for running the U.S. Department of Defense (DoD) acquisition enterprise. This report outlines the results of this research, including comparisons of the costs from these estimates to commercial benchmarks for program management (PM).

Acquisition Functions

A functional decomposition of DoD acquisition and procurement (broadly defined) is developed that can serve as a baseline tool for understanding government functions. Figure S.1 contains a taxonomy of acquisition functions based on our expertise and review of acquisition instructions, guidebooks, and workforce partitioning. As with any portfolio or taxonomy exercise, other approaches can be developed, but this approach aligns well with the common functions as reflected generally in defense organizations, career fields, and acquisition documents.

Estimating Baseline Costs

Given this working taxonomy of the basic government functions in acquisition, how can the DoD measure the cost of executing those functions? *Cost* includes government administration and support contractors in executing the functions illustrated in Figure S.1—not the contractor costs of developing, producing, and sustaining goods and services for the warfighter and other missions. This includes a very wide range of goods and services acquired by the DoD. For example, goods include weapon systems (e.g., aircraft, ships, submarines, land vehicles, software, spacecraft), electronic and communication systems, textiles, weapons, ammunition, and facilities. Contracted services include research and development, knowledge-based services, electronics and communication services, equipment servicing, logistics, transportation, medical services, facilities, and construction (see Under Secretary of Defense for Acquisition, Technology, and Logistics, 2016a, p. 7).

Determining the cost of executing this baseline is challenging because existing accounting and labor-tracking systems are not aligned consistently to isolate and identify all acquisition work. This was recognized in prior efforts (e.g., unpublished 2015 research by RAND colleagues Jeffrey A. Drezner, Irv Blickstein, Mark Arena, Jerry Sollinger, and Charles Nemfakos). Analysis of various options determined that a workforce-based cost estimate—while lacking

Figure S.1
Acquisition Functions

1.0 Program management
    1.1 Business-case and economic analysis
    1.2 Affordability analysis
    1.3 Acquisition strategy
    1.4 Risk management
    1.5 Technical maturity
    1.6 Personnel and team management
    1.7 Business and marketing practices
    1.8 Configuration management

2.0 Research and development

3.0 Engineering
    3.1 Systems engineering
    3.2 Facilities engineering
    3.3 Software/information technology

4.0 Intelligence and security (protection, counterintelligence)
    4.1 Cybersecurity
    4.2 Program protection

5.0 Test and evaluation (T&E)
    5.1 Developmental T&E
    5.2 Operational T&E

6.0 Production, quality, and manufacturing (PQM)

7.0 System and operational issues
    7.1 Spectrum (frequency allocation, emissions, etc.)
    7.2 Environmental
    7.3 Energy

LEGEND
Goods and services
Goods only

8.0 Product support, logistics, and sustainment

9.0 Financial management

10.0 Cost estimating

11.0 Auditing

12.0 Contract administration
    12.1 Contracting actions
    12.2 Contracting strategy
    12.3 Contract peer review
    12.4 Acceptance of deliverables

13.0 Purchasing

14.0 Industrial-base and supply-chain management

15.0 Infrastructure and property management

16.0 Manpower planning and human systems integration

17.0 Training and education
    17.1 Training and education for government execution
    17.2 Training and education for acquired systems

18.0 Disposal

Acquisition Interface Functions

19.0 Requirements: Receive, inform, and fulfill

20.0 Acquisition intelligence: Request, receive, and respond

21.0 Legal counsel: Request and act upon

NOTE: Definitions for each function are provided in the appendix.

in accounting for certain physical infrastructure apart from office space (e.g., test facilities and warehouses)—is the most viable option. Changing the charts of account or implementing activity-based costing are possible alternatives for improving these gaps, but such approaches would likely be extremely costly, time consuming, and politically challenging given their scope and various cultural drivers and motivations behind the current approaches. Further analysis would be warranted before considering a major revision of the accounting charts of account. One of these impediments (which is significant and poses a historic hurdle) is Congress's approach to making appropriations to satisfy its Constitutional prerogatives. Another is the general underlying process that the DoD uses to generate budget estimates from year to year. This requires great fidelity in the approach to making changes in account structures.

Workforce-based costs for the acquisition baseline were estimated using available data, identifying trends since fiscal year (FY) 2008,[2] comparative benchmarks, and other ways to begin gaining perspective on these costs. Multiple approaches and some imprecisions in the available data result in a range of annual estimated costs from about $29 billion to $38 billion in FY 2017 (not counting physical infrastructure). Estimated costs from FY 2008 to FY 2017 are shown in Figure S.2. Trends are generally flat, but because of uncertainties, the trends in total cost range from a decrease of about annual 3.5 percent to an increase of about 6.5 percent. In adjusted (constant) dollars, the estimate for uncertainties ranged from a high of about $42 billion in FY 2011 to a low of about $28 billion in FY 2008. Increases in civil servant costs are generally offset (by chance or design) by decreases in support contractors and IT systems costs over this period (although these may have been a budget adjustment between accounts by the Comptroller, rather than real changes in program-level funding decisions). Data on the job-function distribution of support contractors would greatly reduce the uncertainty range of these estimates.

To gain perspective on these costs and their trends, we identified commercial benchmarks for the amount of PM—a key element in acquisition execution. To try and compare DoD PM

Figure S.2
Estimated Total Acquisition Execution Cost—Civilian, Military, and Support Contractor plus IT Systems (FYs 2008–2017)

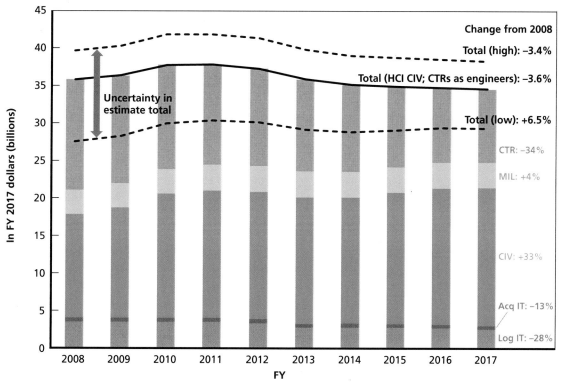

NOTES: Acq IT = acquisition IT; CIV = civil servant; CTR = (support) contractor; HCI = Human Capital Initiative; Log IT = logistics IT; MIL = military (personnel).

---

[2]   The workforce data used as the basis for most of these estimates are available in a relatively consistent form starting in FY 2008. Before that, the Office of the Secretary of Defense's Human Capital Initiative said the data-reporting policies and categories were different, making comparisons of workforce numbers before and after 2008 unsound.

costs with these commercial benchmarks, we totaled the cost of the program and science and technology (S&T) managers in the acquisition workforce.[3] As shown in Figure S.3, PM costs as a function of DoD contracting obligations[4] ran about 1 percent to 1.5 percent in FY 2008 to FY 2017—below the margins of available benchmarks for commercial activity (which ranged from 2 percent to 15 percent across numerous commercial industries). Thus, the amount of PM (including support contractors) appears to be reasonable and perhaps too low to maximize the performance of defense acquisition. It is hard to say conclusively if DoD PM levels are too low because there are differences between government and commercial PM functions and because the commercial benchmarks vary so widely by commercial sector. However, government PM involves significant regulatory processes not found in many commercial sectors. Also, there is evidence of insufficient situational awareness in some DoD programs that breached Nunn-McCurdy cost growth thresholds (see Under Secretary of Defense for Acquisition, Technology, and Logistics, 2016a, p. 28).

Figure S.3
Program and S&T Management Execution Cost Margin and Commercial Benchmarks (FYs 2008–2017)

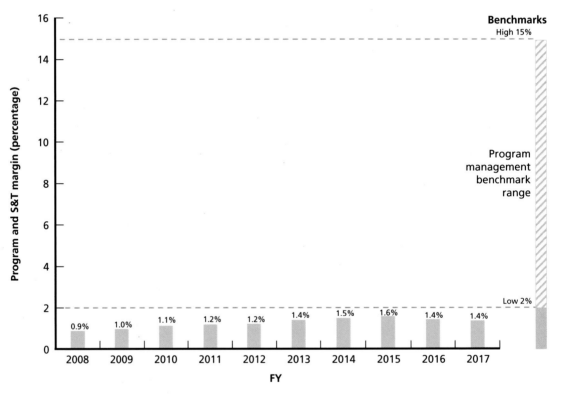

NOTES: Function of total DoD contract obligations and execution costs. Using mean BLS engineer salary for all contractors. Benchmark is Kerzner (1998) (spans other reported benchmark values).
We used the U.S. Bureau of Labor Statistics' engineer salary as the average surrogate for all support contractors for two reasons. First, in our experience, engineers make a large portion of the PM workforce. Second, engineers tend to be at the higher end of the pay scale. Given this, these rates are more conservative toward higher numbers and yet still are below the commercial PM benchmark, making that finding more remarkable. These calculations could be updated if contractor demographic data were available.

---

[3]   We included S&T managers to be as comprehensive as possible and to include early design management.

[4]   *Obligated contract dollars* is a readily available measure of the product of acquisition, which includes research, development, procurement, and sustainment of acquired goods and services.

Rather than contracting dollars, a different perspective is gained by examining cost as a function of the annual number of contracting actions. As a function of contracting transactions (contract awards and modifications), total acquisition workforce costs from FY 2008 to FY 2014 increased from $20,000 to $24,000—about 20 percent. Trends were similar when examining just the contracting workforce costs against transactions ($3,600 to $4,300). Transaction reporting criteria apparently changed between FY 2014 and FY 2015 with a commensurate jump in transactions numbers, so subsequent trends are harder to interpret but seem to have flattened. We did not have the resources to research what led to the increase in transaction numbers, so we just caution that there may have been a change that might make it unsupportable to compare rates before and after the fiscal 2014–2015 boundary.

There are numerous other outputs from the acquisition enterprise beyond the simple measures of contracting dollars and transactions, so these perspectives are limited. Further research can be conducted to examine and control for multiple effects while searching for productivity and efficiency measures.

Finally, while some of these baseline cost measures show increases in some cases, associated benefits may justify such increases to acquisition stakeholders. For example, many (but not all) of DoD's published performance indicators for major defense acquisition programs show recent reductions in cost growth on major defense acquisition programs. While these savings and cost avoidances may be unrelated to some degree, they may be much larger than workforce cost increases. More analysis is needed, but it may indicate that the DoD's and Congress's investments in increasing its acquisition workforce may be showing measurable benefits.

Acknowledgments

We thank our project sponsor, Nancy Spruill, Office of the Under Secretary of Defense (Acquisition and Sustainment) (OUSD[A&S]), for her support throughout this effort. We also thank Philip D. Rodgers and Nicolle Yoder in OUSD(A&S).

We appreciate the support from a number of our colleagues at RAND, including Susan Gates and the Acquisition and Technology Policy Center management team lead by Cynthia Cook, Christopher Mouton, Laura Baldwin, and Joel Predd. Charles Nemfakos and Brian Persons provided very helpful peer reviews. All errors, however, remain those of the authors.

Abbreviations

A&S	Acquisition and Sustainment
ABC	activity-based costing
AWF	acquisition workforce
BLS	U.S. Bureau of Labor Statistics
CAPE	Cost Assessment and Program Evaluation
CIV	civil servant
CTR	(support) contractor
DLA	Defense Logistics Agency
DMDC	Defense Manpower Data Center
DoD	U.S. Department of Defense
DoDI	Department of Defense Instruction
FAR	Federal Acquisition Regulation
FCOM	full cost of manpower
FFRDC	federally funded research and development center
FTE	full-time equivalent
FY	fiscal year
HCI	Human Capital Initiative
IT	information technology
LOB	line of business
MDA	Milestone Decision Authority
MIL	military (personnel)
NDRI	National Defense Research Institute
O&M	Operation and Maintenance
O&S	Operations and Support
OPM	U.S. Office of Personnel Management
OSD	Office of the Secretary of Defense
OUSD(AT&L)	Office of the Under Secretary of Defense (Acquisition, Technology, and Logistics)
PB	President's Budget

PM	program management
RDT&E	Research, Development, Test, and Evaluation
S&T	science and technology
SNaP-IT	Select and Native Programming—Information Technology
T&E	test and evaluation
TIC	total installed cost
USD(AT&L)	Under Secretary of Defense for Acquisition, Technology, and Logistics

CHAPTER ONE

Baselining Defense Acquisition and Procurement

As part of the effort to identify efficiencies in the acquisition system, the Office of the Secretary of Defense (OSD) asked the National Defense Research Institute, a federally funded research and development center (FFRDC) operated by the RAND Corporation, to explore ways for clarifying government execution functions and costs in the acquisition enterprise. This baseline entails listing and defining the basic government acquisition functions, exploring practical data sources for determining the cost of executing these government functions, and testing these data sources to obtain cost estimates for running the U.S. Department of Defense (DoD) acquisition enterprise. This document outlines the results of this research, including comparisons of the costs from these estimates to commercial benchmarks for program management (PM).

Definition of *Acquisition*

Acquisition is defined by the DoD as

> [t]he conceptualization, initiation, design, development, test, contracting, production, deployment, integrated product support (IPS), modification, and disposal of weapons and other systems, supplies, or services (including construction) to satisfy DoD needs, intended for use in, or in support of, military missions. (Defense Acquisition University [DAU], 2017a)

Note that acquisition includes not only procurement but also product support, modification, and disposal of weapon systems as well as the acquisition of other systems, goods, and services. Thus, for this study, we will use the term *acquisition* for simplicity to cover all these activities, including procurement.

Organization of This Report

Chapter Two provides a taxonomy of major government acquisition functions in the DoD. Extensive definitions for each function (citing authoritative sources, when possible) are provided in the appendix. Chapter Three presents available data sources for estimating the cost of executing these functions, including gaps and potential ways for filling these gaps. It then summarizes results from using the best-available data to estimate execution costs and trends from fiscal year (FY) 2008 to FY 2017. Chapter Four provides closing observations and conclusions.

CHAPTER TWO

Acquisition Functions

Just what does the government have to do to acquire goods and services?

This question seems simple at first, but there are many ways in which acquisition is categorized and decomposed. In financial and budgeting terms, the government categorizes dollars into fundamental budget activities:

- Research, Development, Test, and Evaluation (RDT&E)
- Procurement
- Operation and Maintenance (O&M)
- Military Construction and Family Housing
- Military Personnel.

Procurement is clearly acquisition. Much of RDT&E is as well, but it also includes early research referred to as *science and technology* (S&T), which is an enabler of acquisition but normally considered separate. Some of O&M involves acquisition, but O&M also pays for operations well outside of acquisition (such as the operation of military bases). Military construction is included in acquisition, but some family housing is not. Finally, some military personnel perform acquisition, but many do not. Thus, the comptroller budget activities are helpful in some ways for seeing acquisition dollars but are not precisely aligned to acquisition as a partition of DoD activity.

Another way to try and decompose acquisition functions is to look at the major stages in acquisition. These stages relate somewhat with the budget activities but are not perfectly aligned. Table 2.1 breaks out the acquisition functions from the definition in Chapter One by stage. Note that major functions—such as PM, quality assurance, and delivery acceptance—are not included. There is also some overlap with earlier S&T.

Thus, we combined insights from these sources, various guidebooks (DAU, 2002, 2004; Parker, 2011; Marine Corps Systems Command, 2017; Naval Air Systems Command, 2015), acquisition career fields (U.S. Office of Personnel Management [OPM], 2018; Under Secretary of Defense for Acquisition, Technology, and Logistics [USD(AT&L)], 2016b), and our own experience in acquisition to develop a functional taxonomy (Table 2.2). This taxonomy has a number of benefits over budgetary or stage-based functions lists. It allows us to identify functions—such as contracting—that apply across activities and stages. It also allows us to see that while all functions apply to the acquisition of goods (including weapon systems), a number of these functions apply to the acquisition of contracted services as well. Some detailed functions are indented within a hierarchy under other overarching functions; these more-specific functions could be omitted depending on the level of specificity desired, and others could arguably be added.

Definitions for these functions are included in the appendix of this report, along with authoritative sources, when available.

As with any portfolio or taxonomy exercise, other approaches can be developed, but this approach aligns well with the common functions as reflected generally in defense organizations, career fields, and acquisition documents.

Table 2.1
Acquisition Functions, by Stage

Budget Activity	Function
RDT&E	S&T
	Conceptualization
	Initiation
	Contracting
	Design
	Technology maturation and risk reduction
	Engineering and manufacturing development
	Testing
Procurement	Contracting
	Production
	Deployment
	Acquisition strategy
	Testing
Sustainment	Integrated product support and sustainment
	Modification and upgrades
	Testing
Disposal	Disposal

Table 2.2
Acquisition Functions for Goods and Services

Goods	Services	Functions
		Primary Acquisition Functions
X		1.0 Program management
X	X	1.1 Business-case and economic analysis
X	X	1.2 Affordability analysis
X	X	1.3 Acquisition strategy
X	X	1.4 Risk management
X		1.5 Technical maturity
X		1.6 Personnel and team management
X		1.7 Business and marketing practices
X		1.8 Configuration management
X		2.0 Research and development
X		3.0 Engineering
X		3.1 Systems engineering
X		3.2 Facilities engineering
X		3.3 Software/information technology (IT)
X		4.0 Intelligence and security (protection, counterintelligence)
X		4.1 Cybersecurity
X		4.2 Program protection
X		5.0 Test and evaluation (T&E)
X		5.1 Developmental T&E
X		5.2 Operational T&E
X		6.0 Production, quality, and manufacturing (PQM)
X		7.0 System and Operational Issues
X		7.1 Spectrum (frequency allocation, emissions, etc.)
X		7.2 Environmental
X		7.3 Energy
X	X	8.0 Product support, logistics, and sustainment
X	X	9.0 Financial management
X	X	10.0 Cost estimating
X	X	11.0 Auditing
X	X	12.0 Contract administration
X	X	12.1 Contracting actions
X	X	12.2 Contracting strategy

Table 2.2—Continued

Goods	Services	Functions
X	X	12.3 Contract peer review
X	X	12.4 Acceptance of deliverables
X		13.0 Purchasing
X	X	14.0 Industrial-base and supply-chain management
X		15.0 Infrastructure and property management
X		16.0 Manpower planning and human systems integration
X		17.0 Training and education
X		17.1 Training and education for government execution
X		17.2 Training and education for acquired systems
X		18.0 Disposal
		Acquisition Interface Functions
X	X	19.0 Requirements: Receive, inform, and fulfill
X		20.0 Acquisition Intelligence: Request, receive, and respond
X	X	21.0 Legal Counsel: Request and act upon

SOURCES: Author experience and analysis of DAU (2002, 2004), Parker (2011), Marine Corps Systems Command (2017), Naval Air Systems Command (2015), OPM (2018), and USD(AT&L) (2016b).
NOTE: Functions that apply to contracted services as well as goods are shaded in gray for clarity. Supply-chain management generally applies only for goods.

CHAPTER THREE

Estimating Baseline Costs

Now that we have a working taxonomy of the basic government functions in acquisition, the question is how the DoD can measure the cost of executing those functions. These are just the government administrative costs, including their support contractors (CTRs)—not the costs of actually developing, producing, and sustaining goods and services for the warfighter and other missions. This includes a very wide range of goods and services acquired by the DoD. For example, goods include weapon systems (e.g., aircraft, ships, submarines, land vehicles, software, spacecraft), electronic and communication systems, textiles, weapons, ammunition, and facilities. Contracted services include research and development, knowledge-based services, electronics and communication services, equipment servicing, logistics, transportation, medical services, facilities, and construction (see USD[AT&L], 2016a, p. 7).

Not unlike commercial activities, DoD components and the Fourth Estate[1] each have a responsibility to manage and oversee the acquisition of new weapon systems, other goods, and contracted services. These government activities include a range of technical experts, program and financial managers, auditors, contracting officers, senior leadership, and others responsible for executing the defense acquisition enterprise. Below we assess the utility of several different data sources for creating a cost baseline of these activities within the DoD, then test their utility and limitations in cost estimates of the acquisition baseline.

Data Sources for Estimating Baseline Costs

We reviewed the extensive public budget exhibits and justification books published by the DoD and military departments for RDT&E, Procurement, and O&M accounts (Under Secretary of Defense [Comptroller], undated). Unfortunately, acquisition execution costs are rarely tracked in these exhibits by acquisition functions, acquisition organizations, or even by acquisition as a whole. DoD budgets neither align clearly to acquisition functions nor consistently isolate government execution costs from development, production, and sustainment costs by contractors or organic government entities. There are some instances where budget (in dollars or personnel numbers) are available for certain acquisition functions or organizations, but their occurrences are infrequent and inconsistent. We can, of course, easily identify the totals spent on acquisition of RDT&E, procurement, and O&M, but these are not separable

---

[1] Per the DoD, the *Fourth Estate* "is comprised of organizational entities which are not in the Military departments or the Combatant Commands. These include the OSD, the Chairman of the Joint Chiefs of Staff, the Office of the Inspector General, the Defense Agencies, and Field Activities" (Deputy Chief Management Officer, 2015). See also 4th Estate DACM, undated.

between government execution costs and the activities (mostly by contractors) that perform the development, procurement, and sustainment of goods and services. Essentially, DoD's charts of account were established for other historical and organizational reasons than to account for costs by functions such as acquisition. We verified our observation from reviewing the budget exhibits with a former Navy comptroller.

The best option we found was to estimate execution costs by counting and then monetizing the acquisition workforce (AWF). As illustrated in the next section, there are extensive data that align with acquisition functions, but unfortunately, this is an incomplete approach in that certain costs are not accounted for (e.g., some personnel such as engineers who are not formally identified as being in the AWF; activities such as the acquisition of goods and services at military installations; and infrastructure costs such as test equipment).

Relevant Workforce Data for Estimating Enterprise Execution Costs

We identified three primary data sources for AWF:

1. **Human Capital Initiative (HCI) AWF.** This is the official AWF tracking by OSD's HCI (Office of the Under Secretary of Defense for Acquisition and Sustainment, undated). HCI breaks down the workforce into 14 AWF career fields for civil servants (CIV) and military (MIL) (see Tables 3.1 and 3.2). CIV data are published as personnel counts per AWF field—both as a total and by a different set of white-collar career fields designated by OPM (2018, undated[a]). MIL data are published as simple counts without rank or specialty. HCI data are available annually going back to FY 2005. Because of major changes in how the AWF was tracked, HCI considers data before FY 2008 to not be comparable to levels since.

    We were able to obtain CTR total headcount estimates (assumed to be full-time equivalents [FTEs]) from an HCI "data call" (a one-time request for data) to the three military departments in the third quarter of FY 2017 (these data are not part of the officially managed AWF data). The departments of the Navy and Air Force broke out FFRDC counts from other CTRs. These data included historical numbers for FY 2008 or FY 2009, FY 2017 (as of about the third quarter), and future estimates through FY 2022.
2. **FedScope.** This is published CIV workforce data from OPM (undated[b]). This system provides a range of de-identified data for each federal employee, including agency, organization, pay, and OPM career field. Data are readily available going back to FY 2001.
3. **Defense Manpower Data Center (DMDC).** DMDC data include extensive CIV and MIL data. Unfortunately, as we explain in the next subsection, because of major restrictions in working with DMDC data, we only used DMDC data to obtain demographics on the numbers and ranks of MIL officers and enlisted personnel assigned to the AWF.

Relevant Data for Estimating Acquisition Information Systems Costs

Select and Native Programming—Information Technology (SNaP-IT): We used the public exhibit from the President's Budget (PB) request for FY 2018 based on SNaP-IT data.[2]

---

[2]   SNaP-IT is a database application used to plan, coordinate, edit, publish, and disseminate IT budget justification books required by the Congress (see DoD Financial Management Regulation, 2017).

Data Limitations

While the data sources we listed earlier are useful for obtaining an estimate of the government costs of executing the acquisition enterprise, they have limitations. First, as we mentioned earlier, no workforce data align completely with the full range of acquisition execution functions. This difficulty was also recognized in unpublished 2015 research by RAND colleagues Jeffrey A. Drezner, Irv Blickstein, Mark Arena, Jerry Sollinger, and Charles Nemfakos, which tried to correlate AWF and workload (using Unit Identification Codes and contracting data). Acquisition includes not only the development, procurement, and sustainment of major defense weapon systems through formal acquisition programs but also the acquisition and procurement of the myriad goods and services obtained by the DoD. Those efforts involve more than the officially designated acquisition professionals because it includes contributions from other specialists and operators.

Also, much of the underlying sources are the same for the three workforce databases discussed. The HCI data feed DMDC submissions, which in turn feed OPM submissions that FedScope reports on. There are some differences (e.g., there are additional data collected by HCI on the AWF), but the core sets are in common. Thus, while we illustrate how using these three sources provides a range of cost estimates and thus reflects the uncertainties in these estimates, much is in common between them. Still, the different approaches we outline next attempt to compensate for some of the individual limitations in the data sources to see how consistent the results may be.

The HCI data have known gaps and limitations. Only full-time individuals dedicated to certain acquisition functions are included. Also, individual organizations voluntarily opt in or out of the AWF, deciding, for example, whether or not their engineers are part of the AWF based broadly on the organization's primary functions. While these decisions provide insight into those primary organizational functions, there are limits to the precision in these decisions. Some individuals included may not be spending the majority of their time in acquisition functions, while others from organizations opting out of the AWF may actually be performing acquisition functions. In addition, the published HCI AWF data on MIL personnel do not include demographics on rank and the split between enlisted and officers, so other sources were used to determine those demographics. Finally, HCI does not regularly collect contractor FTE estimates. We did obtain estimates by the departments of the Army, Navy, and Air Force from the HCI, but that was a one-time estimate. Moreover, contractor headcount estimates are known to be very difficult and thus less precise because the DoD does not have systems in place to regularly integrate contractor hours spent on a contract.

FedScope personnel data are very helpful because they are available at the individual level (including pay and organization) without the restrictions placed on DMDC data. However, FedScope does not include data on MIL or CTR workforce. Nevertheless, these data are useful for providing a comparison check against an HCI-based CIV estimate. They also allow us to calculate DoD averages for each U.S. Bureau of Labor Statistics (BLS) career field, which are then applied against the HCI demographics data.

DMDC data include extensive CIV and MIL data. Unfortunately, as we indicated earlier, it is much more difficult to gain access to and work with these data because they can include data identifiable to individuals. These restrictions prevent extraction of even de-identified spreadsheets with rows for each person. Nevertheless, we were able to use the detailed AWF data in the DMDC archive to construct rank demographics to apply to the published HCI MIL data.

Finally, some financial management systems in SNaP-IT systems also support acquisition and procurement, but they are not readily coded for acquisition and thus were not included in this estimate. Further analysis could be conducted to assess the description of each system in SNaP-IT to try and identify IT systems costs associated with acquisition.

These types of challenges are not unique to the DoD. The Intelligence Community, for example, has recognized the importance of understanding its workforce and associated costs as part of strategic workforce planning but had to deal with data shortfall and alignment (e.g., tracking of contractor numbers and functions, aligning workforce to activities and missions) (see, for example, Nemfakos et al., 2013).

Future Options for Filling Data Gaps

We briefly examined options for filling execution data gaps in the acquisition line of business (LOB). First, the AWF categories (career fields) could be aligned better with the acquisition functions to some degree, but even if that was done, there is acquisition-related work performed part-time by others normally categorized outside the acquisition functions. Also (as we will see in the next section), there are overlaps between career fields, so we suspect that it would be very difficult to develop a clean partitioning of careers by function without overlap. Also, improvements to the military and civilian workforce career fields would still leave the challenges of tracking CTRs.

Alternatively, the DoD's financial management chart of accounts (e.g., budgeting and funds accounting) could be aligned to LOBs such as acquisition but, as mentioned earlier, there is a long history of alternative organizational reasons why the charts are structured the way they are. Also, different components (e.g., the military departments) have variants in their accounts based on historical and content differences, which would be difficult to reconcile. Such peculiarities and nonuniformity in charts of account are not without precedent; commercial companies design their charts of account for specific reasons and adjust them over time for specific management reasons. The importance of LOB functional tracking would need to be significant to motivate and justify the disruption and cost of this kind of major change to the DoD's accounting. Further analysis would be warranted before considering a major revision of the charts of account.

A different financial management option is to establish activity-based costing (ABC) within existing charts of account that align labor accounting to LOB functions (see, for example, Moore, 2000). This approach would establish time report accounts based on the activity being performed—not just to record hours worked but what function was being done in each hour. Current timecard systems should be able to support ABC accounting by establishing labor charge numbers for LOB functions, but incentives would be needed to ensure that charges are recorded accurately. One way to incentivize reporting would be to align the labor reporting accounts to actual payments (not just large O&M appropriations), but such an approach would greatly increase the amount of changes required. Implementation of ABC in the DoD has been attempted in the past but was not successful, but given the current leadership's emphasis on commercial practices, the DoD might consider implementing ABC. Again, further analysis would be warranted before considering a major revision of the charts of account.

Mapping Acquisition Function to HCI Workforce Career Fields

To better understand how well the existing workforce data align with—and are covered by—the acquisition functions just outlined, we built a table mapping acquisition functions to HCI

AWF career fields (see Table 3.1). Primary contributors are marked by an "X" in the table, while lesser (secondary) contributors are marked with an "o." We also identified areas of known gaps, highlighted in yellow, on the right side of the table; many of these result from nonprogrammatic purchasing (e.g., the acquisition of goods and services by installation personnel in support of base operations) or by personnel who are involved in the acquisition of goods and services on a part-time basis and are not included in the workforce counts.

Notably, many acquisition functions are performed by multiple types of career personnel. For example, configuration management is performed by program managers, engineers, IT personnel, T&E personnel, life-cycle logisticians, and PQM personnel. In addition, operators not coded as AWF also contribute insight (part-time or on assignment) into the benefits of different design configurations, but they are not counted in the AWF numbers.

The mapping shown in this table is not used directly in our cost estimates but rather to see whether the HCI AWF career data spans the acquisition functions identified above. If it did not, then we would be more concerned that major acquisition functions would not be covered by simply monetizing the cost of the AWF. Of course, as noted, there are other people outside the coded AWF that perform acquisition functions (at least part-time), but an AWF cost estimate at least results in measuring a major portion of these costs.

Table 3.1
Mapping Between Acquisition Functions and HCI Workforce Career Fields

Acquisition Functions	HCI AWF Career Categories														Non-AWF		
	Auditing	Business: Cost Estimate	Business: Financial Management	Contracting	Engineering	Facilities Engineering	Information Technology	Life Cycle Logistics	Production, Quality, and Manufacturing	Program Management	Property	Purchasing	S&T Manager	Test and Evaluation	User/Operator	Other	Includes
1.0 Program management/manager										x							
1.1 Business-case and economic analysis	o	x	x	o						x					x	x	Nonprogram purchasing
1.2 Affordability analysis	o	x	x	o						x					x	x	Nonprogram purchasing
1.3 Acquisition strategy				x						x		x					
1.4 Risk management	o	o	o		x		x	x	x	x	o			x			
1.5 Technical maturity					x		x	x	x	x				x			
1.6 Personnel and team management	o	o	o	o	o		o	o	o	x	o	o	o	o	x	x	Nonprogram purchasing
1.7 Business and marketing practices				o						x	o	o	o		x	x	Nonprogram purchasing
1.8 Configuration management					x		x	x	x	x	o		o	x	x	x	Nonprogram purchasing
2.0 Research and development					x		x		o	o			x				
3.0 Engineering					x		x			o							
3.1 Systems engineering					x		x			o							
3.2 Facilities engineering						x				o							
3.3 Software/information technology					x		x			o							
4.0 Intelligence and security (protection, counterintelligence)										o							
4.1 Cybersecurity					x		x			o							
4.2 Program protection					x		x			o							

Table 3.1—Continued

	HCI AWF Career Categories														Non-AWF		
Acquisition Functions	Auditing	Business: Cost Estimate	Business: Financial Management	Contracting	Engineering	Facilities Engineering	Information Technology	Life Cycle Logistics	Production, Quality, and Manufacturing	Program Management	Property	Purchasing	S&T Manager	Test and Evaluation	User/Operator	Other	Includes
5.0 T&E					×		×			o				×			
5.1 Developmental T&E					×		×			o				×			
5.2 Operational T&E					×		×			o				×			
6.0 Production, quality, and manufacturing (PQM)					×				×	o							
7.0 System and operational issues					×					o							
7.1 Spectrum					×					o							
7.2 Environmental					×					o							
7.3 Energy					×					o							
8.0 Product support, logistics, and sustainment					o		o	×		o							
9.0 Financial management			×	o	o					o							
10.0 Cost estimating		×	×	×	o					o							
11.0 Auditing	×		o	o													
12.0 Contract administration	×	×	×	×	o		o	o	o	×	o	×	o	o			
12.1 Contracting actions				×						o							
12.2 Contracting strategy		×		×						×							
12.3 Contract peer review	o			×						o							
12.4 Acceptance of deliverables									×	o		×					

Table 3.1—Continued

Acquisition Functions	HCI AWF Career Categories														Non-AWF		
	Auditing	Business: Cost Estimate	Business: Financial Management	Contracting	Engineering	Facilities Engineering	Information Technology	Life Cycle Logistics	Production, Quality, and Manufacturing	Program Management	Property	Purchasing	S&T Manager	Test and Evaluation	User/Operator	Other	Includes
13.0 Purchasing	o	x	x	x						o		x			x	x	Nonprogram purchasing
14. 0 Industrial-base and supply-chain management										x		x			x	x	Nonprogram purchasing
15.0 Infrastructure and property management	o	o	o	o						o	x				x	x	Nonprogram purchasing
16. Manpower planning and human systems integration					x		x	x		x					x	x	Nonprogram purchasing
17.0 Training and education																	
17.1 For government execution										o						x	Nonprogram purchasing
17.2 For acquired systems					x		x	x		o							
18.0 Disposal	o	o	o	o	o		o	o		o					x	x	Nonprogram purchasing
Acquisition Interface Functions																	
19.0 Requirements: Receive, inform, and fulfill					x		x	x		x							
20.0 Acquisition intelligence: Request, receive, and respond					x		x			x							
21.0 Legal counsel: Request and act upon	x	x	x	x						x					x		

SOURCES: Author experience and analysis of DAU (2002, 2004); Parker (2011), Marine Corps Systems Command (2015), OPM (2018), Naval Air Systems Command (2017), and USD(AT&L) (2016b).

NOTES: Functions on the left-most column that generally only apply to the acquisition of goods are colored white, while those that apply to both goods and services are shaded purple. Personnel contributions to a function that are deemed primary (major) are marked with an "X," while lesser contributions are marked by an "o." Gaps are highlighted in yellow.

Workforce-Based Cost Estimates

Using the workforce data just described, we generated two estimates for the cost of executing the defense acquisition enterprise. The first estimate is predominantly HCI AWF based, while the second uses a CIV estimate from FedScope (see Table 3.2).

Cost Estimate for FY 2017

Figure 3.1 shows the results of applying these workforce data to estimate the acquisition baseline costs for FY 2017. These estimates range from a low of about $29 billion to a high of about $38 billion (about 30 percent higher).

On the low side, the HCI-based CIV estimate comes to $18.4 billion with a MIL estimate of $3.4 billion. The HCI CTR data separated FFRDCs from other contractors for the Navy and Air Force. Assuming the Army uses about as many FFRDC FTE as the Navy (i.e., fewer than the Air Force) and assuming a salary of about $150,000, we get a total FFRDC cost for acquisition of about $1 billion. The FFRDC totals for acquisition from this estimate are about half of the total DoD obligations for FFRDCs. For example, the Congressional Research Service reported that obligations to DoD FFRDCs totaled $1.9 billion in FY 2015 (then-year dollars). Finally, the low-end estimate for the remaining CTR labor is about $3.5 billion, using the BLS salaries for office, production, and administrative support (BLS occupation codes 43-0000 and 51-0000; see Figure 3.1).

On the high side, the FedScope-based cost estimate for CIV is $20.1 billion—about 9 percent higher than the HCI-based estimate. The MIL costs are the same as before. For the

Table 3.2
Workforce-Based Cost Estimate Approaches

	Number (Source)	Dollars per Person (Source)
HCI AWF based		
CIV	OPM career field count per AWF career field (HCI overview)	DoD actual averages by OPM career field (FedScope) + 30% CAPE FCOM
MIL	# per AWF career field (HCI overview)	(DoD average per rank) + 40% CAPE FCOM
CTR	Army, Navy, and Air Force: provided for third quarter of FY 2017 Fourth Estate: Assume same fraction as CIV+MIL	Used BLS salary averages (various ones could apply) + 100% fringe + 15% profit
FedScope based		
CIV	X + Y, where X = number of DoD CIV in certain acquisition-dominated OPM career fields Y = number of DoD CIV in certain OPM career fields that are in certain acquisition-dominated organizations	Per-person salary (FedScope) + 30% CAPE FCOM
MIL CTR	Same as HCI-based estimate above.	

NOTES: CAPE FCOM = Cost Assessment and Program Evaluation Full Cost of Manpower. Department of Defense Instruction (DoDI) 7041.04 (2013) defines FCOM to reflect fringe benefits and other nonlabor costs, such as general and administrative; overhead; facilities; maintenance; repair; insurance; office supplies; printing; rent; security; and support services, travel, and utilities. It is intended to enable the comparison of the full cost of CIV, MIL, and CTR labor.

Figure 3.1
Cost Estimates of the Acquisition Baseline (FY 2017)

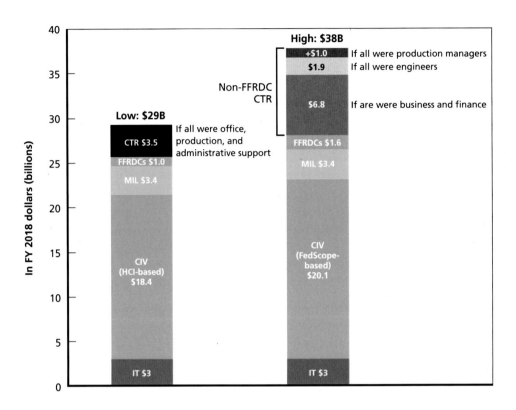

high estimate, we assume the Army FFRDC levels are the same as for the Air Force, increasing the estimate to about $1.6 billion. For the non-FFRDC CTR costs, we show three ranges. If all were providing business and financial operations (BLS 13-0000), the cost would be $6.8 billion instead of $3.5 billion in the low estimate. If instead they were all engineers (BLS 17-2000), then it would be $1.9 billion higher. Finally, if all CTRs were production managers, then it would be another $1 billion higher. We provide these ranges of estimates because we do not have job or salary demographics for the CTRs in the data provided to OSD HCI. It is possible that the military departments have cost data for CTRs (from which they estimated FTE); if so, then future baseline estimates could use those data and reduce the uncertainty in the estimates.

Cost-Estimating Details

To estimate CIV costs from the HCI AWF headcounts, we extracted tables that showed for each HCI career field the numbers of civilians by OPM career fields multiplied by the FedScope cost average for those OPM career fields (see, for example, Table 3.3). An additional 30 percent was added to the salaries to reflect the FCOM.[3]

For MIL costs, we applied DoD averages for individual ranks against the overall AWF rank demographics from DMDC multiplied by the total number of MIL personnel for each

---

[3]   OSD's Director of Cost Analysis and Performance Assessment defines FCOM in DoDI 7041.04 (2013) to reflect fringe benefits and other nonlabor costs, such as general and administrative; overhead; facilities; maintenance; repair; insurance; office supplies; printing; rent; security; and support services, travel, and utilities.

Table 3.3
Composition and Size of the HCI AWF Program Management Career Field, by OPM Occupational Series (Third Quarter of FY 2017)

OPM Occupational Series	Count	Percentage
1101—Business and Industry Specialist	3,461	28
0340—Program Manager	3,185	26
0343—Management and Program Analyst	2,932	24
0301—Administration and Program Staff	1,334	11
0801—Engineer, General	683	5.6
2210—Information Technology Management	157	1.3
0855—Engineer, Electronics	120	0.97
1515—Operations Research Analyst	52	0.42
0830—Engineer, Mechanical	30	0.24
0802—Engineering Technician	23	0.19
Other	336	2.7
Total civilian count	**12,313**	**100**

SOURCE: HCI, 2017.

AWF career field. An additional 40 percent CAPE FCOM was then used to reflect fringe costs.

Finally, we used the third-quarter FY 2017 CTR FTE counts for the departments of the Army, Navy, and Air Force and applied various BLS salary averages (with 100 percent fringe and 15 percent profit) to obtain ranges of costs. Since HCI did not have CTR counts for the fourth estate, we resorted to assuming that these numbers were in the same proportion to service CTRs as the CIV plus MIL counts (even though we recognize the differences in manning between the military departments and the Fourth Estate).[4] That is, about 16 percent of the CIV and MIL AWF total are in the Fourth Estate, so we assume about 16 percent of total support CTRs are also in the Fourth Estate and scaled the numbers accordingly. If additional data on CTR support numbers in the Fourth Estate were to become available, then the estimate could be readily updated to reflect any inaccuracies from this assumption. The 100 percent fringe ratio is based on our past experience in working with contractor cost data.

An Alternative Approach to Compare with the HCI Estimate

To provide a separate check on the HCI-based CIV estimate, we applied a combined organization and career-field method to the FedScope data. First, we assumed that all DoD CIV with certain acquisition-dominant OPM career fields (see Table 3.4) are working acquisition and were included in the estimate using the FedScope salary data for that individual plus 30 percent for CAPE FCOM. Second, we added other individuals (and their 30 percent CAPE FCOM) if they are coded in certain other acquisition-related OPM career fields (e.g.,

---

[4] We realize that this assumption may be weak, but we did not have other data to estimate the size of the contractor workforce in the Fourth Estate. We know it is not zero, and we might think that the use of contractor support is somewhat similar across the DoD in practice. Again, better data would help here.

Table 3.4
OPM Career Fields Assumed to Be Always Performing Acquisition Tasks

Program Management (0340)	Industrial Specialist (1150)
Management and Program Analysis (0343)	Production Control (1152)
Logistics Management (0346)	Quality Assurance (1910)
Contact Representative (0962)	Supply Program Management (2003)
Contracting (1102)	Supply Clerical and Technician Series (2005)
Purchasing (1105)	
Procurement Clerical and Technician (1106)	

NOTE: For some perspective, currently there are 24,946 Contracting officials (1102s) in the DoD.

0801 General Engineering; see full list in Table 3.5) and work in acquisition-related organizations (e.g., the Army Material Command; see full list in Table 3.6). Since FedScope has no MIL and CTR data, we used the same cost estimates from the HCI-based estimate noted earlier for the FedScope estimate. We understand that we are making some (perhaps overly simplified) assumptions on the similarity of data and organizational structures of the DoD. For example, not all people of a certain career field in a certain organization perform (or do not perform) acquisition. Also, our selection of career fields to include may miss some or include others (e.g., it is unclear how many 0560s [Budget Analysis Series] support acquisition compared with other budgetary activities, yet we included them). Nevertheless, this is an initial attempt meant to be used to compare with the AWF-based estimate, and this alternative approach could be refined by further research or informed by further discussions with personnel experts in the DoD.

The results of estimating CIV costs based on organizations and career fields was about 9 percent higher than the HCI AWF–based estimate ($20.1 billion instead of $18.4 billion, respectively, for FY 2017). Thus, using the AWF as a source appears to be reasonable given the lack of data on acquisition execution costs and associated cost estimation difficulties.

Figure 3.2 shows the relevant BLS occupations for DoD support CTRs and their average (mean) annual wages as reported in FY 2016 dollars (BLS, 2017). We do not have demographic information on the distribution of this kind of contractor support to the acquisition enterprise, so we use these to generate a range of possible CTR costs. FFRDC average salary levels are not reported, so we used a rough approximation of $150,000; OSD has actual fully loaded cost per FTE for DoD FFRDCs, which could then be used in future cost estimates to refine this value.

Baseline Costs Trends

The HCI AWF data are reported back to FY 2008, so we used the same analytic approach to show AWF cost estimates for FY 2008 through FY 2017 in constant FY 2017 dollars (i.e., adjusted for inflation) (see Figure 3.3). Nominal estimates are shown as stacked bars for CIV (blue), MIL (green), and CTR (gray) labor costs. The percentages to the right of the plot indicate the change from FY 2008 to FY 2017. The nominal cost for CIV is the HCI AWF estimate, and the nominal CTR values use the BLS Engineering salary (BLS, 2017) for all CTRs, including FFRDCs. CIV, MIL, and CTR salary levels were deflated back to FY 2008 using the deflators for DoD CIV, DoD MIL, and U.S. Consumer Price Index for Urban Wage Earners and Clerical Workers (CPI-W), respectively. The CTR data from the military departments in FY 2017 included intermittent years going back to FY 2008 or 2009; we used a linear interpolation between the years provided (except for the FY 2008 Navy numbers, for which we used their FY 2009 values to avoid the risk of artificially inflating the values beyond any reported).

Table 3.5
OPM Career Fields Assumed to Be Performing Acquisition Tasks If Their Organization Is Predominantly Acquisition Related

0110	Economist Series	0899	Engineering and Architecture Student Trainee Series	1603	Equipment, Facilities, and Services Assistance Series
0132	Intelligence Series				
0140	Workforce Research and Analysis Series	1082	Writing and Editing Series	1670	Equipment Services Series
0142	Workforce Development Series	1083	Technical Writing and Editing Series	1701	General Education and Training Series
0201	Human Resources Management Series	1103	Industrial Property Management Series	1702	Education and Training Technician Series
0203	Human Resources Assistance Series	1107	Property Disposal Clerical and Technician Series	1710	Education and Vocational Training Series
0301	Miscellaneous Administration and Program Series			1712	Training Instruction Series
		1130	Public Utilities Specialist Series	1720	Education Program Series
0341	Administrative Officer Series			1730	Education Research Series
		1160	Financial Analysis Series	1740	Education Services Series
0342	Support Services Administration Series	1202	Patent Technician Series	1750	Instructional Systems Series
		1210	Copyright Series		
0501	Financial Administration and Program Series	1301	General Physical Science Series	1801	General Inspection, Investigation, Enforcement, and Compliance Series
0503	Financial Clerical and Technician Series	1306	Health Physics Series		
		1310	Physics Series	1802	Compliance Inspection and Support Series
0505	Financial Management Series	1311	Physical Science Technician Series	1805	Investigative Analysis Series
0510	Accounting Series	1313	Geophysics Series		
0530	Cash Processing Series	1315	Hydrology Series	1810	General Investigation Series
0540	Voucher Examining Series	1316	Hydrologic Technician Series		
				1815	Air Safety Investigating Series
0560	Budget Analysis Series	1320	Chemistry Series		
0561	Budget Clerical and Assistance Series	1321	Metallurgy Series	1822	Mine Safety and Health Inspection Series
		1330	Astronomy and Space Science Series		
0801	General Engineering Series			1825	Aviation Safety Series
		1340	Meteorology Series	1849	Wage and Hour Investigation Series
0802	Engineering Technical Series	1341	Meteorological Technician Series		
				1850	Agricultural Warehouse Inspection Series
0803	Safety Engineering Series	1350	Geology Series		
		1360	Oceanography Series	1860	Equal Opportunity Investigation Series
0806	Materials Engineering Series	1361	Navigational Information Series		
				1862	Consumer Safety Inspection Series
0808	Architecture Series	1370	Cartography Series		
0809	Construction Control Technical Series	1371	Cartographic Technician Series	1863	Food Inspection Series
				1889	Import Compliance Series
0810	Civil Engineering Series	1372	Geodesy Series	2001	General Supply Series
0817	Survey Technical Series	1373	Land Surveying Series	2010	Inventory Management Series
0819	Environmental Engineering Series	1374	Geodetic Technician Series		
				2030	Distribution Facilities and Storage Management Series
0828	Construction Analyst Series	1380	Forest Products Technology Series		
0830	Mechanical Engineering Series	1382	Food Technology Series	2032	Packaging Series
		1384	Textile Technology Series	2091	Sales Store Clerical Series
0840	Nuclear Engineering Series			2099	Supply Student Trainee Series
		1386	Photographic Technology Series		
0850	Electrical Engineering Series			2101	Transportation Specialist Series
0854	Computer Engineering Series	1397	Document Analysis Series		
				2102	Transportation Clerk and Assistant Series
0855	Electronics Engineering Series	1399	Physical Science Student Trainee Series		
				2110	Transportation Industry Analysis Series
0856	Electronics Technical Series	1501	General Mathematics and Statistics Series		
				2131	Freight Rate Series
0858	Bioengineering and Biomedical Engineering Series	1510	Actuarial Science Series	2135	Transportation Loss and Damage Claims Examining Series
		1515	Operations Research Series		
		1520	Mathematics Series	2144	Cargo Scheduling Series
0861	Aerospace Engineering Series	1521	Mathematics Technician Series	2150	Transportation Operations Series
		1529	Mathematical Statistics Series	2151	Dispatching Series
0871	Naval Architecture Series			2152	Air Traffic Control Series

Table 3.5—Continued

0873	Marine Survey Technical Series	1530	Statistics Series	2154	Air Traffic Assistance Series
0880	Mining Engineering Series	1531	Statistical Assistant Series	2161	Marine Cargo Series
0881	Petroleum Engineering Series	1540	Cryptography Series	2181	Aircraft Operation Series
		1541	Cryptanalysis Series	2183	Air Navigation Series
0890	Agricultural Engineering Series	1550	Computer Science Series	2185	Aircrew Technician Series
0893	Chemical Engineering Series	1599	Mathematics and Statistics Student Trainee Series	2199	Transportation Student Trainee Series
0895	Industrial Engineering Technical Series	1601	Equipment, Facilities, and Services Series	2210	Information Technology Management Series
0896	Industrial Engineering Series			2299	Information Technology Student Trainee Series

NOTE: Organizations deemed predominantly acquisition related are listed in Table 3.6.

Table 3.6
Organizations Assumed to Be Predominantly Acquisition Related

AF03	Air Force Operational Test and Evaluation Center
AF1M	Air Force Materiel Command
AF1S	Headquarters, Air Force Space Command
AF2A	Air Force Cost Analysis Agency
AF2L	Air Force Technical Applications Center
AF2R	Air Force Program Executive Office
ARAT	U.S. Army Test and Evaluation Command
ARSC	U.S. Army Space and Missile Defense Command/U. S. Army Forces Strategic Command
ARX1	U.S. Army Material Command
ARX6	U.S. Army Aviation and Missile Command
ARX7	U.S. Army Tank Automotive and Armament Command (Tacom)
ARX8	U.S. Army Communications Electronics Command
ARXB	U.S. Army Chemical Materials Activity
ARXC	U.S. Army Sustainment Command
ARXD	U.S. Army Contracting Command
ARXK	Materiel Acquisition Activities
ARXP	U.S. Army Security Assistance Command
ARXQ	U.S. Army Joint Munitions Command
ARXR	U.S. Army Research, Development and Engineering Command
ARXT	U.S. Army Military Surface Deployment and Distribution Command
ARXX	Materiel Readiness Activities
DD07	Defense Logistics Agency
DD10	Defense Contract Audit Agency
DD13	Defense Advanced Research Projects Agency
DD27	Missile Defense Agency
DD63	Defense Contract Management Agency
DD68	Department of Defense Test Resource Management Center
DD71	Defense Microelectronics Activity
DD81	Defense Acquisition University
DD82	National Reconnaissance Office
NV19	Naval Air Systems Command
NV23	Naval Supply Systems Command
NV24	Naval Sea Systems Command
NV30	Strategic Systems Programs Office
NV39	Space and Naval Warfare Systems Command

NOTE: This list may differ somewhat from the self-identified organizations in the HCI AWF system.

Nominal CIV costs for the AWF increased 33 percent, and MIL costs increased 4 percent. These increases are not surprising because Congress and the DoD made overt efforts to rebuild the AWF since their decline in the 1990s (USD[AT&L], 2016b). Conversely, nominal CTR costs dropped 34 percent since FY 2008, bringing the net total change decrease since FY 2008 to –0.7 percent. In other words, the nominal estimate shows that the AWF costs have been essentially flat, with the DoD essentially insourcing the work of CTRs.

Figure 3.2
Average Annual Wages by BLS Occupation Series (FY 2016)

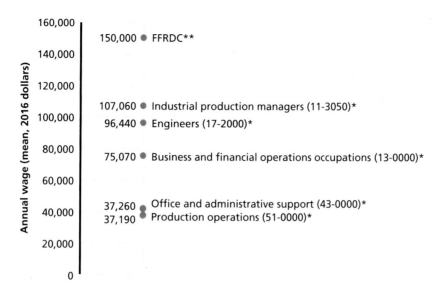

SOURCE: BLS, 2017.
NOTES: The average annual wage across the FFRDCs is known to OSD but not
available to RAND. * = from BLS May 2016 survey, ** actual value to be determined.

Figure 3.3
**Estimated Total Acquisition Execution Cost—Civilian, Military, and Support Contractor plus IT
Systems (FYs 2008–2017)**

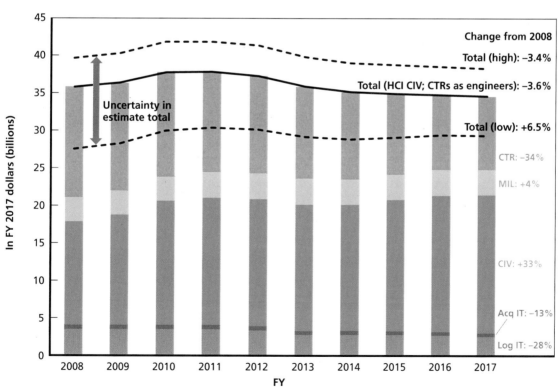

NOTES: Acq IT = acquisition IT; Log IT = logistics IT.

As with Figure 3.1, we explored the uncertainty in the CIV estimate by using an organizational and OPM career-based algorithm on FedScope data. We also explored the uncertainty in the CTR data using different assumptions of CTR demographics. The lower dashed line shows the low estimate in these excursions. The low estimate uses the HCI AWF estimate for CIV, the same MIL estimate, the lower FFRDC estimate (assuming the Army uses the same number of FFRDC FTEs as the Navy), and the low BLS career field (administrative support and production—BLS 43-0000 and 51-0000). At this low end of the estimates, the net total change from FY 2008 would be an increase of 6.5 percent (i.e., the less-expensive reductions in contractor costs would not quite compensate for the increased CIV numbers).

For the high estimate, we increased the yearly HCI AWF CIV cost estimate by 9 percent, which is the amount by which the FedScope cost estimate was higher than the HCI-based estimate in FY 2017. We also used the higher FFRDC count estimate, assuming the Army used the same number of FFRDC FTE as reported by the Air Force. For the balance of the CTRs, we used the high BLS salary for production managers (BLS 11-3050). The net effect at these high estimate levels is a –3.4 percent change (decrease).

Thus, as a result of estimating costs using different approaches, we found the uncertainty band ranges from a 6.5 percent increase to a 3.5 percent decrease, with the actual value probably lying somewhere between these bounds. These values demonstrate the importance of including changes in CTRs when understanding execution cost trends. Otherwise, the AWF costs would appear to have increased significantly. It also shows that better demographic data (or actual costs) are needed to reduce the uncertainty in costing the acquisition baseline.

For IT system costs from FY 2011 to FY 2017, we use SNaP-IT exhibits from PB 2013, PB 2015, PB 2018, PB 2018, and PB 2019. Exhibits are not published earlier, so we used FY 2011 values for FYs 2008–2010. Alternatively, we could have used the decreasing trend to estimate higher values for FYs 2008–2010, but that might have projected larger budget decreases than is factually evident.

To further examine the changes in CTRs, we examined how those changes differ by component. Figure 3.4 shows the reported changes in support CTRs by the three military departments. Values between reported years were interpolated. As before, the Fourth Estate contractor levels were scaled to be equivalent to have the same ration that the CIV plus MIL HCI AWF levels are between the military departments and the Fourth Estate.

In comparison to the prior chart on CIV changes, the Army and Navy departments had much larger decreases in percentage than the Air Force since FY 2008 even though the Army actually had a decrease in CIV costs since FY 2008. Thus, the Amy apparently had not been insourcing work while the Air Force and Navy have been (in net).

Figure 3.5 breaks out the HCI-based CIV cost estimate from the previous chart by major component. Interestingly, the Army's changes since FY 2008 have been slightly negative, while the Navy and Air Force have increased significantly. Of the three departments, the Air Force's changes were the highest, but they also started from the smallest base and remain smaller than the two peer departments.

Figure 3.6 breaks out the earlier HCI-based MIL cost estimate in Figure 3.3 by major component. In contrast to the size of its CIV workforce cost, the Air Force MIL AWF is the largest among the four services. The Air Force and Navy have been essentially flat since FY 2008. The U.S. Marine Corps (USMC) and Army increased by 33 and 15 percent, respectively, but are much smaller in absolute terms.

Figure 3.4
Number of Acquisition Support Contractors (FYs 2008–2017)

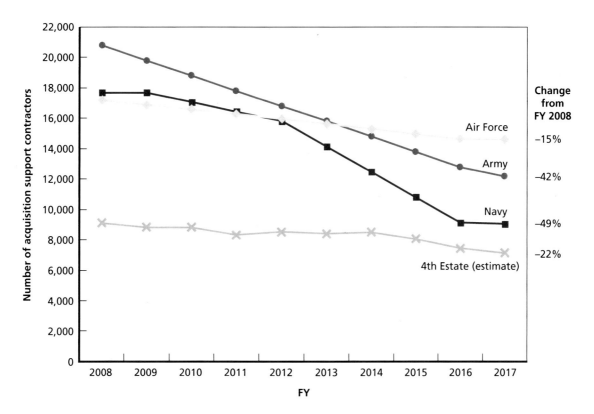

Figure 3.5
Estimated Civilian Acquisition Workforce Cost by Component (FYs 2008–2017)

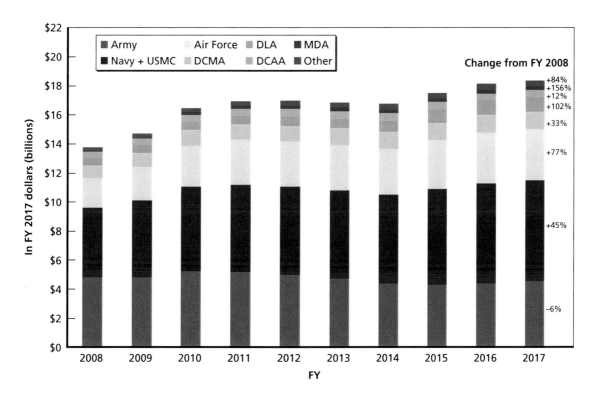

Figure 3.6
Military AWF Costs by Component (FYs 2008–2017)

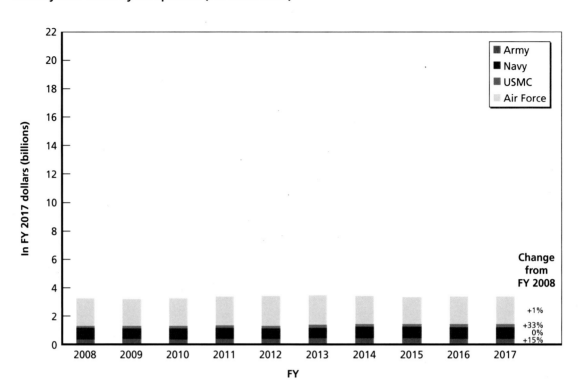

Figure 3.7 breaks out the HCI-based CIV and MIL cost estimates by HCI career field. Engineering and contracting are the largest career field, followed by life-cycle logistics and PM. All AWF career fields except the business–financial management have grown since FY 2008.

Baseline Costs as a Function of Output

Now that we have cost estimates (both in absolute terms and in trends since FY 2008) for executing the government acquisition enterprise (including CTRs), we need some perspective about whether these costs are reasonable, low, or excessive.

Cost Margin as a Function of Contracting Obligations

One approach is to calculate the cost margin[5] relative to total activity, then compare that margin to commercial benchmarks. If we use annual contract obligations as the output, then the cost margin is represented by the following equation:

$$Cost\ margin = \frac{Execution\ cost}{Contracted\ obligations\ +\ CIV\ and\ MIL\ execution\ costs}.$$

---

[5]   Margins are the portion of total spending or revenue going toward the activity of interest (i.e., as opposed to a markup, which is a percentage added to the producing operation to pay for the activity of interest).

Figure 3.7
Civilian and Military AWF Costs by HCI Career Field (FYs 2008–2017)

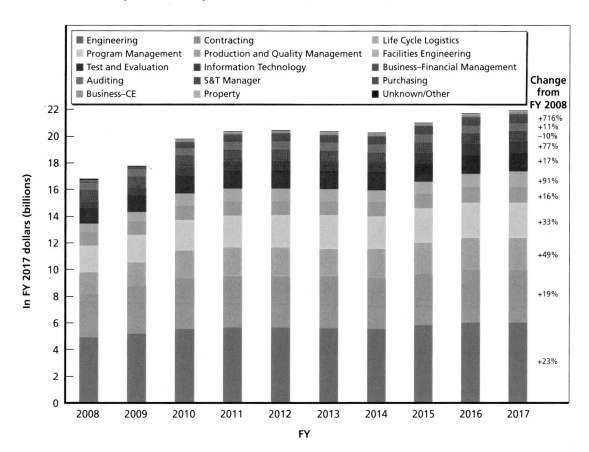

Note that since contractor support costs are already included in contract obligations, we only add CIV and MIL execution costs to the total DoD contract obligations for the denominator.

Total Cost Margin

In FY 2017, DoD contract obligations totaled $320.2 billion, as reported by the Federal Procurement Data System–Next Generation. Thus, the execution cost margin ranges from 8.5 percent ($29.3 billion / $345 billion) to 11 percent ($38.3 billion / $347.2 billion):

$$Total\ cost\ margin = \frac{\$29.3B}{\$345B}\ to\ \frac{\$38.3B}{\$347.2B} = 8.5\%\ to\ 11\%\ .$$

Using the same approach going back to FY 2008 yields margins ranging from 7.5 percent to 11.6 percent (see Figure 3.8).

Figure 3.8
Total Acquisition Enterprise Execution Cost Margin (FYs 2008–2017)

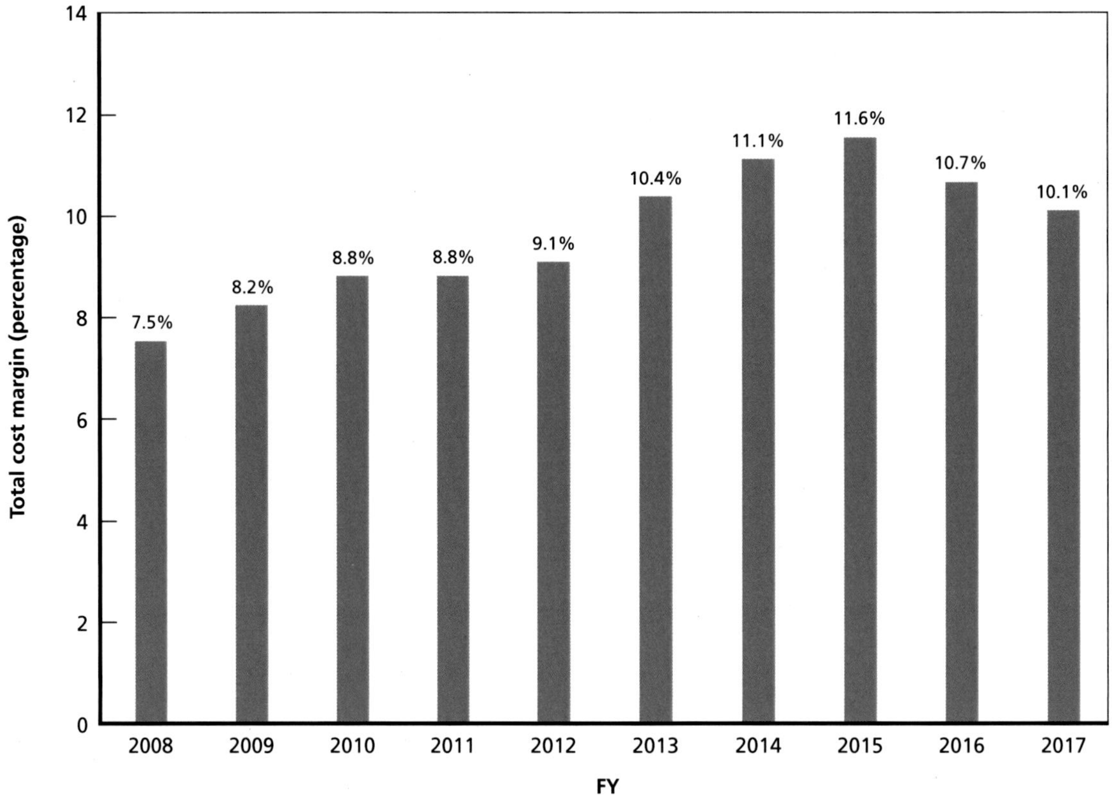

NOTE: Function of total DoD contract obligations.

Program Management Cost Margin

If we just consider the PM portion of the baseline AWF (to include S&T management), the FY 2017 margin is about 1.4 percent (when using the HCI AWF-based CIV estimate and the BLS engineering salary [BLS, 2017] for all CTRs and FFRDCs):

$$PM\ cost\ margin = \frac{\$4.6B}{\$342B} = 1.4\% \ .$$

The values since FY 2008 range from 0.9 to 1.6 percent (see Figure 3.9).

Commercial Benchmarks for Execution Cost Margins

While cost margins provide some perspective, they do not allow decisionmakers to judge whether they are too high, too low, or about right. Ideally, we would compare these cost margins against benchmarks (especially in other contexts, such as the commercial sector, where cost efficiency is balanced against benefits) to help determine whether baseline investments are set at the appropriate levels. Unfortunately, there is no clear commercial analogy to total government execution costs (even if we ignore the differences between government and commercial regulatory requirements); the problem is that acquisition baseline includes the full range of

Figure 3.9
Program and S&T Management Execution Cost Margin and Commercial Benchmarks (FYs 2008–2017)

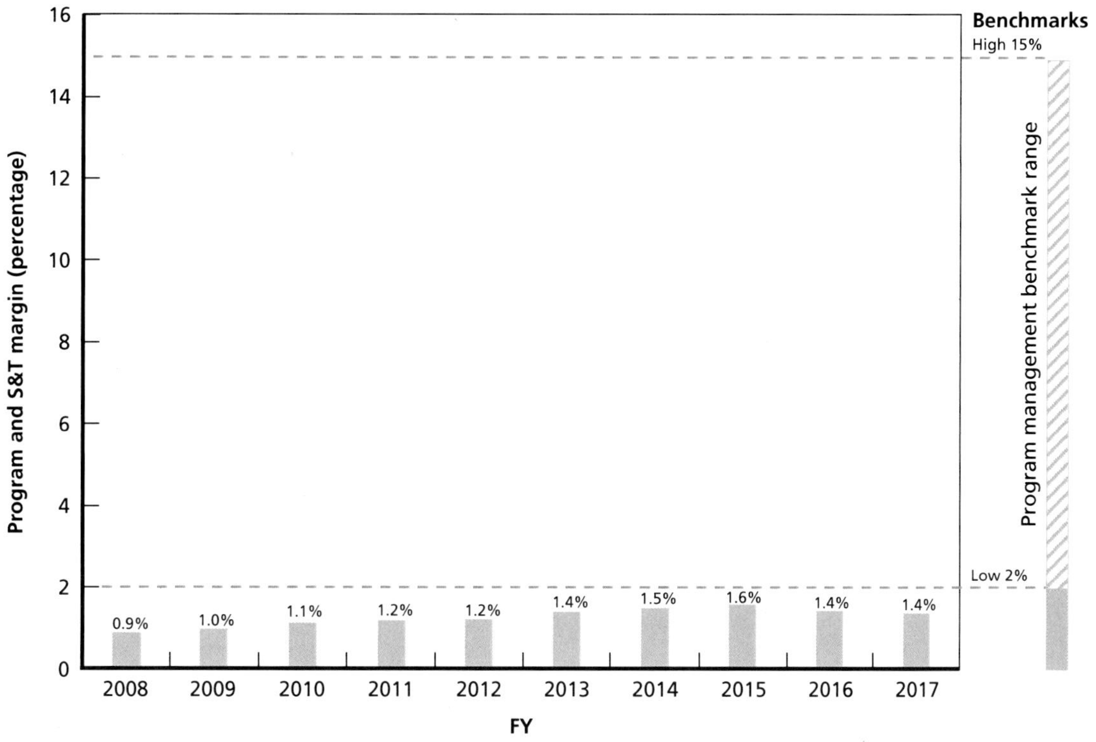

NOTES: Function of total DoD contract obligations + execution costs. Using mean BLS engineer salary for all contractors. Benchmark is Kerzner (1998) (spans other reported benchmark values).
We used the BLS engineer salary (BLS, 2017) as the average surrogate for all CTRs for two reasons. First, in our experience, engineers make a large portion of the PM workforce. Second, engineers tend to be at the higher end of the pay scale, so, if anything, these rates are more conservative toward higher numbers yet are still below the commercial PM benchmark, making that finding more remarkable. These calculations could be updated if contractor demographic data were available.

acquisition functions, such as testing and systems engineering, which, in the private sector, are included as operational activity instead of overhead.

Fortunately, however, there exist a number of reported benchmarks for commercial PM. Research covering a wide range of commercial industries found that PM cost margins[6] range from **2 percent to 15 percent**, varying with project size, industry, and the degree of project controls needed (Heywood and Allen, 1996; Ibbs and Kwak, 1998a, 1998b; Kerzner, 1998; Byrne, 1999; D'Alessandro, 1999; Taylor, 2008; Easton, 2010, 2011). Note that these sources sometimes distinguish between overall *project management* and a subset of *project controls*. Assuming they are being consistent in their terminology, this introduces some caution in reviewing their benchmark values (i.e., which are overall PM and just for the project controls subset). These distinctions may be arising because different industries have different needs for the various functions that PM provides. In construction, for example, there may be less planning done by the PM than in a research and development project.

---

[6]   PM cost margins are the portion of total project costs going toward PM (i.e., as opposed to a markup, which is a percentage of non-PM costs).

Figure 3.9 compares the DoD's PM (and S&T manager) cost margins since FY 2008 to the commercial PM benchmark.

Commercial PM Benchmark Details

Ibbs and Kwak (1998a, 1998b) report that 21 out of 38 surveyed companies of varying types reported that the cost of PM services relative to total project cost **averaged about 6 percent**.[7] Those invited to respond included 15 engineering-construction companies, ten information management and movement companies, ten information systems companies, and three high-tech manufacturing companies, including AT&T, Bechtel, Chevron, East Bay Municipal Utility District, Eichleay Engineers, Hewlett-Packard, IBM, Kodak, Lucent Technologies, Proctor & Gamble, and Sun Microsystems.

D'Alessandro (1999) asserts that small- to medium-sized commercial architecture and engineering projects less than about $1.5 million (adjusted to FY 2017 dollars[8]) should generally spend **10 percent** on PM, and larger projects should spend between **5 percent and 10 percent**.

Byrne (1999) reports on PM in industrial construction. While he asserts that "as much as is necessary and no more" is the correct answer to how much should be spent on PM, he provides some quantitative guidelines. PM cost margins should range from **7 percent to 11 percent**. On **larger projects** (which are probably more analogous to DoD programs), project controls specialists are added (e.g., for estimating, scheduling, and cost control), raising the PM costs to **9–15 percent**.[9]

Kerzner (1998), in a seminal textbook in multiple editions on project management, says that PM margins should range between **2 percent and 15 percent**.

Data published by Jones (1997, p. 196) show that the average management effort on software development ranges from **11 percent to 13 percent** across different domains (management information systems, outsourced, commercial, and systems). Software management averages 16 percent for military software because of added standards and oversight. It increases from a low of 10 percent for very small efforts with 10 function points to 17 percent for software with 100,000 function points.

For the project controls subset of PM, Heywood and Allen (1996) reported cost margins for project controls ranging from 1 percent to 5 percent to "significantly improve the likelihood of project success." This range is similar to what Byrne (1999) reported three years later for the project controls portion of PM for larger projects. Their data came from surveying peers across multiple (unidentified) industries (including, at least, major commercial construction projects) on projects mostly in the $1 million–$230 million range (and one at $1 billion). When plotting margin versus total installed cost (TIC), it showed a rough curve fit, showing how margins decrease exponentially with increasing TIC, asymptoting at just under 1 percent for billion-dollar projects. Taylor (2008) and Easton (2010, 2011) asserted that a lower margin would lead to the program manager becoming overwhelmed by critical tasks.

---

[7]   Mean is 6.4 percent, median 5.1 percent, standard deviation 5.1 percent, minimum 0.3 percent, maximum 16.6 percent.

[8]   Using the BLS U.S. Consumer Price Index Inflation Calculator ("CPI Inflation Calculator," undated).

[9]   Byrne (1999) reports that the PM portion of the TIC of a project is about 7.2–11.2 percent (1.7 percent during conceptualization and design, plus 1.5 percent of the TIC during engineering, plus 4–8 percent of TIC during construction, citing Mahoney (1990).

Anecdotally, a former program manager who ran large, multimillion-dollar research and development programs at multiple major defense prime contractors in the late 1980s to early 2000s said the general rule was always to keep PM costs at or below **10 percent** of the total value of the programs. This anecdote is supported by D'Alessandro (1999).

Generally, the larger the project, the smaller the PM cost margins should be (Heywood and Allen, 1996; Byrne, 1999), and the ranges will vary by industry (Byrne, 1999).

While these data give reference ranges for PM margin benchmarks in the commercial sector across various industries, they do not control for or identify how high the margin should be in specific industries and when projects involve significant research and development. Nevertheless, these demonstrate that, based on commercial standards, the PM margins seen in the DoD appear at least reasonable and perhaps too low to maximize benefits. It is hard to say conclusively if DoD PM levels are too low because there are differences between government and commercial PM functions and because the commercial benchmarks vary so widely and recognize that functions vary greatly by area. However, government PM involves significant regulatory processes not found in many commercial sectors. Also, there is evidence of insufficient situational awareness in some DoD programs that breached Nunn-McCurdy cost growth thresholds (see USD[AT&L], 2016a, p. 28).

With respect to the benefits from PM, Heywood and Allen (1996) asserted that project controls, strategy, and planning have been quantified as adding 5–20 percent in dollar value (probably in the construction industry that they work in). They did not, however, cite their sources.

Costs per Contract Transactions

Another potentially useful (but difficult to execute rigorously) perspective on baseline costs could come from the average annual costs per contract transaction. Since goods and services are acquired through contracts, the number of contracting actions (transactions) could help reflect the ups and downs of acquisition and whether baseline costs are adjusted based on this activity. Simply counting transactions is overly simplistic because it does not control for the very wide variance in the size and execution difficulty of different DoD contracts. Also, any modification to a contract counts as a transaction no matter how minor it may be. Nevertheless, it could be a start toward seeing if contracting activity influences the acquisition baseline. Future analysis is needed to examine whether more sophisticated controls for transaction variables could adjust for these limitations.

Figure 3.10 illustrates some of these challenges and the potential. It plots the total acquisition cost from the baseline (using the nominal engineering BLS salary [BLS, 2017] for all contractors and the lower HCI AWF CIV estimate plus IT costs) divided by the number of contract transactions each fiscal year as reported by USAspending.gov (the green line). Contracting transactions are but one of many outputs from acquisition, but it illustrates the concept. A more sophisticated approach for future analysis might be to test for correlation with multiple activity and output variables to determine how much workforce changes may be due to activity changes and how much appear to be absolute growths for other reasons.

We see that costs per transaction rose about 16 percent from FY 2008 to FY 2014 (R-squared of 0.88). From FY 2015 to FY 2017, it dropped, but this may be an artifact of the transaction numbers for FY 2015 still adjusting to new reporting criteria. We separated the data into these two time periods because there was a dramatic increase in reported transactions between FY 2014 and FY 2015 because of adjustments by the Defense Logis-

Figure 3.10
Total Acquisition Execution Costs per Contract Transaction (FYs 2008–2017)

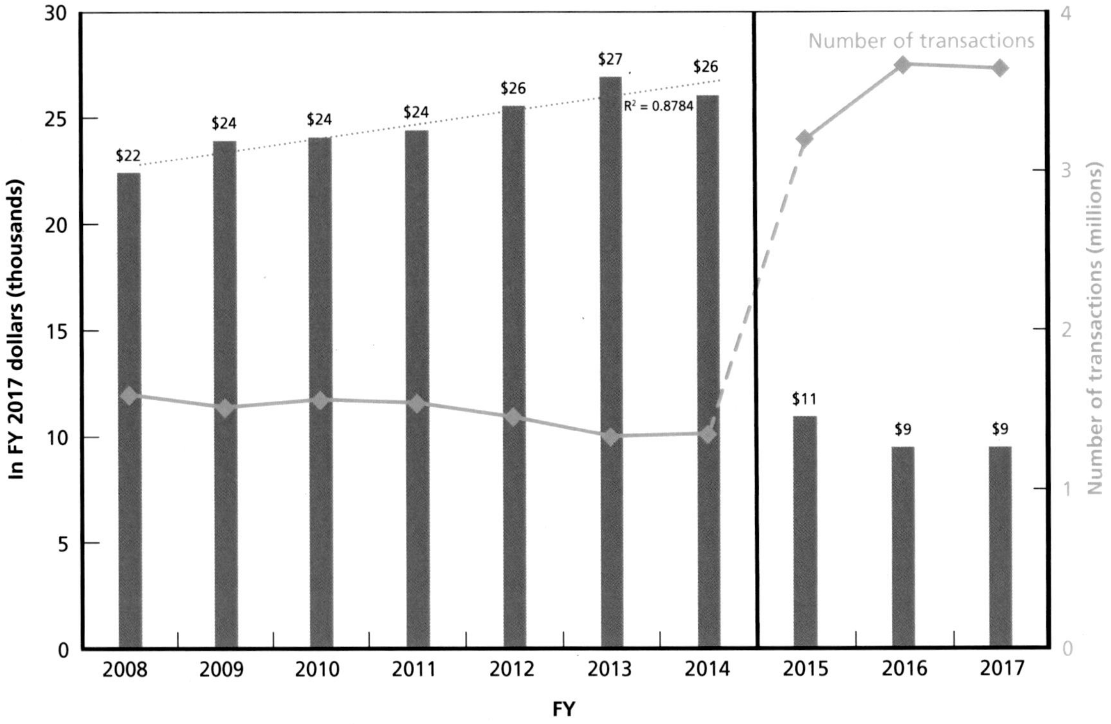

SOURCE: Transaction counts from USAspending.gov, undated.
NOTES: Using mean BLS engineering for all contractors and FFRDC. Baseline cost per transaction (blue bars) are on the y-axis on the left. Transaction values are on the green y-axis on the right. Annual transactions counts were summary totals extracted from USAspending.gov on January 17, 2018. They do not appear to include the very high volume of lowdollar transactions from USTRANSCOM. The increase in reported transactions between FY 2014 and FY 2015 are because of adjustments by the DLA to more fully comply with reporting policies.

tics Agency (DLA) to more fully comply with reporting policies. These data from USAspending.com also do not appear to include the very high volume of low-dollar transactions from U.S. Transportation Command, again illustrating the complications from carefully performing this kind of analysis (i.e., deciding what should be included in the transaction count and how to adjust for wide variation in the work difficulty between transactions).

Figure 3.11 plots an estimate of the contracting workforce costs (CIV, MIL, and a proportional fraction of CTR and IT costs[10]) per contract transaction (using the same y-axis scale as in Figure 3.10 to facilitate comparison). Again, we split the data into two periods given the jump in transactions from FY 2014 to FY 2015 because of DLA reporting changes. The results are similar in shape as in Figure 3.10 on total costs, but they are scaled down—increasing about 16 percent from FY 2008 to FY 2014 (R-squared of 0.78). Similarly, we do not know if the decrease from FY 2015 is an artifact of the transaction reporting changes.

[10] CIV and MIL contracting labor costs constitute about 20 percent of the total AWF costs, so we included that percentage of the CTR and IT cost totals in the contracting costs. Of course, the estimate of CTR costs to support contracting could be refined with more specific data on CTRs by acquisition function supported. Also, further detailed analysis of individual IT elements in SNaP-IT could refine which elements (and their costs) align to contracting functions.

Figure 3.11
Contracting (CIV, MIL, CTR) Execution Costs per Contract Transaction (FYs 2008–2017)

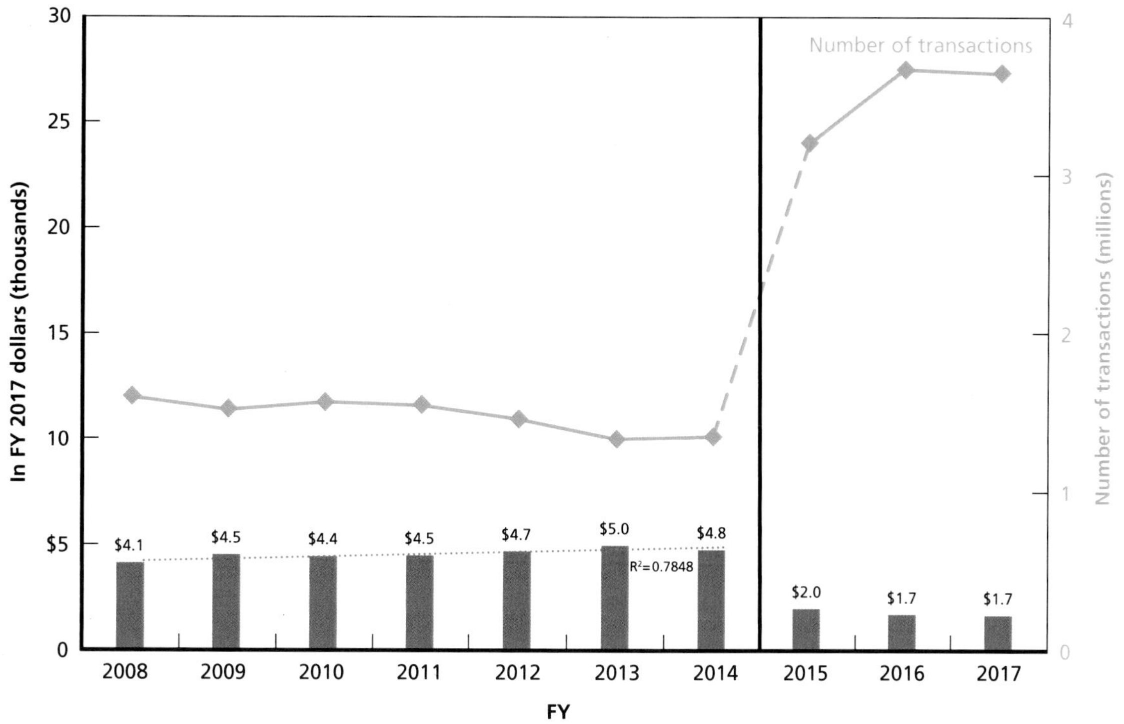

SOURCE: Transaction counts from USAspending.gov, undated.
NOTES: Using mean BLS engineering for all contractors and FFRDC. Contracting workforce cost per transaction (brown bars) are on the y-axis on the left. Transaction values are on the green y-axis on the right. Annual transactions counts were summary totals extracted from USAspending.gov on January 17, 2018. They do not appear to include the very high volume of low-dollar transactions from USTRANSCOM. The increase in reported transactions between FY 2014 and 2015 are because of adjustments by the DLA to more fully comply with reporting policies.

Broader Performance Measures

Finally, we note that other performance factors may provide insights into benefits ensuing from any increases in the acquisition baseline. For example, OSD has been tracking institutional performance on major defense acquisition programs using various measures (see USD[AT&L], 2013, 2014, 2015, 2016a). Many measures, such as the one shown in Figure 3.12, show recent improvements or flattening while others show decreased performance. Similarly, some measures of the acquisition baseline costs show increases while others are flat, uncertain, or decreasing trends. It is very difficult analytically to associate, say, improvements in a performance measure with increased baseline investments based solely on contemporaneous arguments. However, there have been statistically significant improvements coinciding with overt investments by Congress and the DoD to improve and increase the AWF. Without at least qualitatively considering the possibility of a relationship, management may merely seek to eliminate any cost increases when trends are present.

Figure 3.12
Five-Year Moving Average of Annual Growth of Earned Value Contract Costs (FYs 2008–2017)

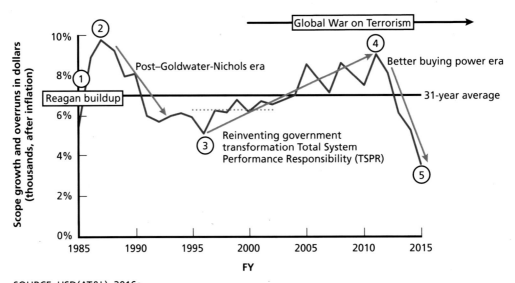

SOURCE: USD(AT&L), 2016a.
NOTES: The numbered circles indicate apparent transition points identified by the DoD. The dashed line indicates part of period three where the DoD indicated the data may have been flat.

CHAPTER FOUR

Conclusion

This report provides a starting point for baselining the government acquisition enterprise. The functional decomposition covers a moderate range of essential functions while trying to keep the list a manageable size. Others can be constructed, but this list provides a useful working point for future refinement.

Understanding and measuring the cost of executing the government functions remains challenging because DoD accounting and budgeting systems are not designed to separate acquisition functions from other activities.[1] Similar difficulties (and the importance of strategic planning and management) faced by the Intelligence Community when it tried to understand and rebuild its workforce capabilities (Nemfakos et al., 2013). Data could be aligned (e.g., by establishing ABC), but such an effort would be very costly, time-consuming, and could run counter to the drivers that established the current charts of account in the first place.

As shown in this report, workforce-based cost estimating augmented by IT infrastructure cost data is a viable alternative, although they lack insight into certain physical infrastructure apart from certain infrastructure (e.g., test facilities and warehouses) because of the accounting limitations (discussed in this report). Since experience indicates that workforce is a dominant contributor to acquisition execution costs, we postulate that a workforce-based cost estimate provides a reasonable and useful tool. By using two different approaches, we obtain estimates for CIV workforce costs that were within about 10 percent of one another—a reasonable starting point for future work. Better CTR and FFRDC data (both in numbers and demographics) would be helpful, and CTR cost data (instead of numbers) would, of course, be even better. Still, these approaches give a rough sense of the cost magnitude and uncertainty range of running the defense acquisition enterprise as long as precision is not overly assumed. More data and more learning would certainly help.

Once we have a cost estimate, the challenge is determining whether those costs are reasonable and beneficial. Comparing costs with external benchmarks (such as the commercial PM benchmarks) indicates that costs appear reasonable or perhaps a bit low for optimum benefit. Examining costs as a function of outputs is also useful, but again, without benchmarks or benefit-cost analysis, it is hard to determine whether these levels are reasonable. Trend analysis helps, but structural changes in output measures such as contract transaction numbers make it hard to compare results over long periods. Further research could examine and control for multiple effects while searching for productivity and efficiency measures.

---

[1] This difficulty was also recognized in unpublished 2015 research by RAND colleagues Jeffrey A. Drezner, Irv Blickstein, Mark Arena, Jerry Sollinger, and Charles Nemfakos.

Finally, while some of these baseline cost measures show increases in some cases, associated benefits may justify such increases to acquisition stakeholders. For example, many (but not all) of the DoD's published performance indicators for major defense acquisition programs show recent reductions in cost growth on major defense acquisition programs. While these savings and cost avoidances may be unrelated to some degree, they may be much larger that workforce cost increases. More analysis is needed, but it may indicate that the DoD's and Congress' investments in increasing its AWF may be showing measurable benefits.

Acquisition Baseline Definitions

Working Definitions of Acquisition and Procurement

This appendix outlines the official definitions for acquisition and procurement. Both are from the online *DAU Glossary* (2017a), which is referred to by DoDI 5000.02 (2018) for definitions of terms in the defense acquisition system.

> Acquisition—The conceptualization, initiation, design, development, test, contracting, production, deployment, integrated product support (IPS), modification, and disposal of weapons and other systems, supplies, or services (including construction) to satisfy DoD needs, intended for use in, or in support of, military missions.

> Procurement—Act of buying goods and services for the government.

> Note that *procurement* is a subset (or equivalence) of acquisition, so the decompositional details in the acquisition function were used below.

Acquisition Functional Decomposition

For the purpose of understanding the costs of executing the defense acquisition enterprise, there is no single best functional decomposition of defense acquisition. The DoD could use the elements in the definition of *acquisition* (see above), but some common functions, such as PM, are not explicitly called out, and these functions do not readily align to with available cost or personnel tracking data.

Alternatively, the DoD could use personnel career fields as the acquisition functions, but government career fields do not always align well with acquisition functions. Some common functions are conducted by generalists, some by specialists, and others by functional specialists, higher-level specialists (such as program managers). Nevertheless, there is reasonable alignment with some career-based categorizations—especially the list of AWF functions defined by the HCI. Figure A.1 contains such a list of basic government functions involved in acquisition and procurement. This list was created using the HCI AWF list and augmenting it with our analysis of major functions based on our expertise in acquisition, along with reviews of selected acquisition guidebooks (DAU, 2002, 2004; Parker, 2011; Naval Air Systems Command, 2015; Marine Corps Systems Command, 2017). Other lists could be constructed, but this list provides a reasonable balance between length and depth while including many of the common major functions cited in the PM guidebooks and from our experience.

Figure A.1
Acquisition Functions

1.0 Program management
  1.1 Business-case and economic analysis
  1.2 Affordability analysis
  1.3 Acquisition strategy
  1.4 Risk management
  1.5 Technical maturity
  1.6 Personnel and team management
  1.7 Business and marketing practices
  1.8 Configuration management

2.0 Research and development

3.0 Engineering
  3.1 Systems engineering
  3.2 Facilities engineering
  3.3 Software/information technology

4.0 Intelligence and security (protection, counterintelligence)
  4.1 Cybersecurity
  4.2 Program protection

5.0 Test and evaluation (T&E)
  5.1 Developmental T&E
  5.2 Operational T&E

6.0 Production, quality, and manufacturing (PQM)

7.0 System and operational issues
  7.1 Spectrum (frequency allocation, emissions, etc.)
  7.2 Environmental
  7.3 Energy

LEGEND
| Goods and services |

Goods only

8.0 Product support, logistics, and sustainment

9.0 Financial management

10.0 Cost estimating

11.0 Auditing

12.0 Contract administration
  12.1 Contracting actions
  12.2 Contracting strategy
  12.3 Contract peer review
  12.4 Acceptance of deliverables

13.0 Purchasing

14.0 Industrial-base and supply-chain management

15.0 Infrastructure and property management

16.0 Manpower planning and human systems integration

17.0 Training and education
  17.1 Training and education for government execution
  17.2 Training and education for acquired systems

18.0 Disposal

Acquisition Interface Functions

19.0 Requirements: Receive, inform, and fulfill

20.0 Acquisition intelligence: Request, receive, and respond

21.0 Legal counsel: Request and act upon

Definitions of Primary Acquisition Functions

Below are definitions for the acquisition functions shown in Figure A.1.

1.0. Program Management/Manager

PM is an overarching acquisition function that incorporates a wide range of acquisition functions. The scope of a specific program may be limited in authority to focus on near-term development and production or may incorporate the full life cycle of the acquired system. DAU (2017a) defines *PM* as follows:

> *Program management.* The process whereby a single leader exercises centralized authority and responsibility for planning, organizing, staffing, controlling, and leading the combined efforts of participating/assigned civilian and military personnel and organizations,

for the management of a specific defense acquisition program or programs, throughout the system life cycle.

Its definition of the individual responsible for PM gives some added insight into what PM entails:

> *Program manager:* Designated individual with responsibility for and authority to accomplish program objectives for development, production, and sustainment to meet the user's operational needs. The program manager shall be accountable for credible cost, schedule, and performance reporting to the Milestone Decision Authority (MDA).

1.1. Business-Case and Economic Analysis

DAU provides the following definition of *economic analysis*:

> 1.1.1. *Economic analysis (EA):* A systematic approach for evaluating the costs of a program or a set of alternatives. An EA evaluates the relative economic (financial) costs of different technical alternatives, design solutions, and/or acquisition strategies, and provides the means for identifying and documenting their costs. Normally associated with Automated Information System (AIS) acquisition programs, it is a statutory requirement for Major Automated Information Systems (MAIS).

It does not provide a definition of business-case analysis, but it includes a broader view of the importance of the good or service to be acquired from a larger mission perspective of the DoD. An economic analysis provides insights into the costs and alternatives, but the larger arguments for or against a certain acquisition would be contained in the business case.

1.2. Affordability Analysis

DAU provides the following definition of *affordability analysis*. An extensive description of affordability analysis is provided in Enclosure 8 of the current DoDI 5000.02 (2018).

> *Affordability analysis:* Long-range planning and decision making that determines the resources a Component can allocate for each new capability by ensuring that the total of all such allocations—together with all other fiscal demands that compete for resources in the Component—are not above the Component's future total budget projection for each year. (DAU, 2017a)

1.3. Acquisition Strategy

> 1.3.1. *Acquisition strategy (AS):* Describes the Program Manager's plan to achieve program execution and programmatic goals across the entire program life cycle. Summarizes the overall approach to acquiring the capability (to include the program schedule, structure, risks, funding, and the business strategy). Contains sufficient detail to allow senior leadership and the Milestone Decision Authority (MDA) to assess whether the strategy makes good business sense, effectively implements laws and policies, and reflects management's priorities. Once approved by the MDA, the Acquisition Strategy provides a basis for more detailed planning. The strategy evolves over time and should continuously reflect the current status and desired goals of the program. (DAU, 2017a)

1.4. Risk Management

1.4.1. *Risk management*: A five-step iterative process to plan, identify, analyze, mitigate, and monitor program risks. (DAU, 2017a; for additional information about risk management, see DoD, 2017)

1.5. Technical Maturity

1.5.1. *Technology readiness level (TRL):* One level on a scale of one to nine, e.g., "TRL 3," signifying active research and development has been initiated. Pioneered by the National Aeronautics and Space Administration (NASA), adapted by the Air Force Research Laboratory (AFRL), and adopted by the Department of Defense as a method of estimating technology maturity during the acquisition process. The lower the level of the technology at the time it is included in a product development program, the higher the risk that it will cause problems in subsequent product development. (DAU, 2017a; for additional information, see DoD, 2009)

1.5.2. *Technical risk:* The risk that arises from activities related to technology, design and engineering, manufacturing, and the critical technical processes of test, production, and logistics. (DAU, 2017a)

1.6. Personnel and Team Management

1.6.1. *Manpower scheduling and loading:* Effective and efficient utilization and scheduling of available manpower according to individual skills to ensure required manufacturing operations are properly coordinated and executed. (DAU, 2017a)

1.6.2. *Manpower:* The total supply of persons available and fitted for service. Indexed by requirements including jobs lists, slots, or billets characterized by descriptions of the required people to fill them. (DAU, 2017a)

1.7. Business and Marketing Practices

1.7.1. *Market research:* A process for gathering data on product characteristics, suppliers' capabilities, and the business practices that surround them. Includes the analysis of that data to inform acquisition decisions. There are two types of market research, strategic market research and tactical market research. (DAU, 2017a)

1.7.2. *Strategic market research:* Includes all the activities that acquisition personnel perform continuously to keep themselves abreast of technology and product developments in their areas of expertise. (DAU, 2017a)

1.7.3. *Tactical market research:* A phase of market research conducted in response to a specific materiel need or need for services. (DAU, 2017a)

1.8. Configuration Management

1.8.1. *Configuration management (CM):* A management process for establishing and maintaining consistency of a product's performance, functional, and physical attributes with its requirements, design and operational information throughout its life. (DAU, 2017a; for additional information, see DoD, 2001)

2.0. Research and Development

2.1. *Research, development, test, and evaluation (RDT&E):* Activities for the development of a new system, or to expand the performance of fielded systems. (DAU, 2017a)

2.2. *Development:* The process of working out and extending the theoretical, practical, and useful applications of a basic design, idea, or scientific discovery. Design, building, modification, or improvement of the prototype of a vehicle, engine, instrument, or the like as determined by the basic idea or concept. Includes all efforts directed toward programs being engineered for Service use that have not yet been approved for procurement or operation, and all efforts directed toward development engineering and test of systems, support programs, vehicles, and weapons that have been approved for production and Service deployment. (DAU, 2017a, 2017b)

3.0. Engineering
3.1. Systems Engineering

3.1.1. *Systems engineering:* An interdisciplinary, methodical, and disciplined approach for the specification, design, development, realization, technical management, operations, and retirement of a system. (DAU, 2017a; for additional information, see DAU 2017b and International Organization for Standardization, 2010)

3.2. Facilities Engineering

3.2.1. *Facilities and infrastructure:* One of the 12 Integrated Product Support (IPS) Elements. It encompasses the permanent and semi-permanent real property assets required to support a system, including studies to define types of facilities or facility improvements, location, space needs, environmental and security requirements, and equipment. It also includes facilities for training, equipment storage, maintenance, supply storage, ammunition storage, and so forth. The objective of this IPS Element is to identify, plan, resource, and acquire facilities to enable training, maintenance and storage to maximize the effectiveness of system operation and the logistics support system at the lowest total ownership cost. (DAU, 2017a; for additional information, see DoD, 2016)

3.3. Software and Information Technology Engineering

3.3.1. *Computer software (or software):* Computer programs, procedures, and possibly associated documentation and data pertaining to the operation of a computer system. (DAU, 2017a)

3.3.2 *Information technology (IT):* Any equipment or interconnected system or subsystem of equipment, used in the automatic acquisition, storage, analysis, evaluation, manipulation, management, movement, control, display, switching, interchange, transmission, or reception of data or information by the executive agency, if the equipment is used by the executive agency directly, or is used by a contractor under a contract with the executive agency that requires the use of: (1) that equipment, (2) that equipment to a significant extent in the performance of a service or the furnishing of a product. It includes computers, ancillary equipment (including imaging peripherals, input, output, and storage devices necessary for security and surveillance), peripheral equipment designed to be controlled by a central processing unit of a computer, software, firmware and similar procedures, services (also support services), and related resources. It does not include any equipment acquired by a federal contractor incidental to a federal contract. (DAU, 2017a)

4.0 Intelligence and Security (protection, CI)

4.0.1. *Counterintelligence (CI):* Information gathered and activities conducted to identify, deceive, exploit, disrupt, or protect against espionage, other intelligence activities, sabotage, or assassinations conducted for or on behalf of foreign powers, organizations or persons or their agents, or international terrorist organizations or activities. (DoD, 2019)

4.0.2. *Security:* 1. Measures taken by a military unit, activity, or installation to protect itself against all acts designed to, or which may, impair its effectiveness. 2. A condition that results from the establishment and maintenance of protective measures that ensure a state of inviolability from hostile acts or influences. 3. With respect to classified matter, the condition that prevents unauthorized persons from having access to official information that is safeguarded in the interests of national security. (DoD, 2019)

4.0.3. *Protection:* Preservation of the effectiveness and survivability of mission-related military and nonmilitary personnel, equipment, facilities, information, and infrastructure deployed or located within or outside the boundaries of a given operational area. (DoD, 2019)

4.1 Cybersecurity

4.1.1. *Cybersecurity:* Prevention of damage to, protection of, and restoration of computers, electronic communications systems, electronic communication services, wire communication, and electronic communication, including information contained therein, to ensure its availability, integrity, authentication, confidentiality, and nonrepudiation. (DAU, 2017a; for additional information see DoDI 8500.01, 2014)

4.2 Program Protection

4.1.2. *Program Protection:* The integrating process for managing risks to DoD warfighting capability from foreign intelligence collection; from hardware, software, and cyber vulnerability or supply chain exploitation; and from battlefield loss throughout the system life cycle. (DAU, 2017a; for additional information, see DoDI 5000.02, 2018)

5.0. Test and Evaluation (T&E)

5.1. Developmental T&E

5.1. *Developmental test and evaluation (DT&E):* (1) Any testing used to assist in the development and maturation of products, product elements, or manufacturing or support processes. (2) Any engineering-type test used to verify status of technical progress, verify that design risks are minimized, substantiate achievement of contract technical performance, and certify readiness for initial operational testing. Developmental tests generally require instrumentation and measurements and are accomplished by engineers, technicians, or soldier operator-maintainer test personnel in a controlled environment to facilitate failure analysis. (DAU, 2017a)

5.2. Operational T&E

5.2. *Operational test and evaluation (OT&E):* The field test, under realistic conditions, of any item (or key component) of weapons, equipment, or munitions for the purpose of determining the effectiveness and suitability of the weapons, equipment, or munitions for use in combat by typical military users; and the evaluation of the results of such tests. (DAU, 2017a)

6.0. Production, Quality, and Manufacturing (PQM)

6.0.1. *Procurement:* Act of buying goods and services for the government. (DAU, 2017a)

6.0.2. *Quality:* The composite of materiel attributes including performance features and characteristics of a production or service to satisfy a customer's given need. (DAU, 2017a)

6.0.3. *Quality assurance (QA):* A planned and systematic pattern of all actions necessary to provide confidence that adequate technical requirements are established, that products and services conform to established technical requirements, and that satisfactory performance is achieved. (DAU, 2017a)

6.0.4. *Quality control (QC):* The system or procedure used to check product quality throughout the acquisition process. (DAU, 2017a)

6.0.5. *Quality audit:* A systematic examination of the acts and decisions with respect to quality in order to independently verify or evaluate the operational requirements of the quality program or the specification or contract requirements for a product or service. (DAU, 2017a)

6.0.6. *Manufacturing:* The process of making an item using machinery, often on a large scale, and with division of labor. (DAU, 2017a)

6.0.7. *Manufacturing engineering:* That specialty of professional engineering which applies engineering procedures to manufacturing processes and methods of production of industrial commodities and products. It requires the ability to plan the practices of manufacturing, to research and develop the tools, processes, machines and equipment, and to integrate the facilities and systems for producing quality products with optimal expenditure. Used in conjunction with design engineering and other functional engineering specialties to create a producible design, that is, a design that can be easily and economically produced. (DAU, 2017a)

7.0. System and Operational Issues
7.1. Spectrum (Frequency Allocation, Emissions, etc.)

7.1.1. *Spectrum supportability risk assessment (SSRA):* Risk assessment performed by DoD Components for all Spectrum Dependent (S-D) systems to identify risks as early as possible and to affect design and procurement decisions accordingly. These risks are reviewed at acquisition milestones and are managed throughout the system's lifecycle. (DAU, 2017a)

7.1.2. *Frequency allocation application (DD Form 1494):* Certification by the National Telecommunication and Information Administration (NTIA) that a candidate system conforms to the spectrum allocation scheme of the United States and its possessions. Requirements for obtaining spectrum support for new telecommunications systems, or major modifications of an existing system, are found in the NTIA Manual of Regulations and Procedures for Federal Radio Frequency Management. Some host nations have similar certifications but requirements vary. (DAU, 2017a)

7.1.3. *Electromagnetic environmental effects (E3).* The impact of the electromagnetic environment (EME) upon the operational capability of military forces, equipment, systems, and platforms. E3 encompasses the electromagnetic effects addressed by the disciplines of electromagnetic compatibility (EMC), electromagnetic interference (EMI), electromagnetic vulnerability (EMV), electromagnetic pulse (EMP), electronic protection (EP), electrostatic discharge (ESD), and hazards of electromagnetic radiation to personnel (HERP), ordnance (HERO), and volatile materials (HERF). E3 includes the electromagnetic effects generated by all EME contributors including radio frequency (RF) systems, ultra-wideband devices, high-power microwave (HPM) systems, lightning, precipitation static, etc. (DAU, 2017a; for additional information, see DoD, 2010)

7.2. Environmental

7.2.1. *Environmental assessment (EA):* Contains an estimate of whether a proposed system will adversely affect the environment or be environmentally controversial, in which case an Environmental Impact Statement (EIS) is prepared. (DAU, 2017a)

7.3. Energy

7.3.1. *Operational energy:* Energy required for training, moving, and sustaining military forces and weapons platforms for military operations. (DoD, 2019)

8.0. Product Support, Logistics, and Sustainment

8.0.1. *Product support (PS):* The package of support functions required to field and maintain the readiness and operational capability of major weapon systems, subsystems, and components, including all functions related to weapon system readiness. (DAU, 2017a)

8.0.2. *Integrated product support (IPS):* A key life cycle management enabler, IPS is the package of support functions required to deploy and maintain the readiness and operational capability of major weapon systems, subsystems, and components, including all functions related to weapon systems readiness. The package of product support functions related to weapon system readiness which can be performed by both public and private entities, includes the tasks that are associated with the Integrated Product Support (IPS) Elements which scope product support. (DAU, 2017a)

8.0.3. *Logistics:* Planning and executing the movement and support of forces (DAU Glossary).

8.0.4. *Acquisition logistics:* Technical and management activities conducted to ensure supportability implications are considered early and throughout the acquisition process to minimize support costs and to provide the user with the resources to sustain the system in the field. (DAU, 2017a)

8.0.5. *Life cycle logistics:* Translates force provider capability and performance requirements into tailored product support to achieve specified and evolving life cycle product support availability, reliability, and affordability parameters. Includes life cycle product support planning and execution, seamlessly spanning a system's entire life cycle, from Materiel Solution Analysis (MSA) to disposal. (DAU, 2017a)

8.0.6. *Life cycle sustainment:* Translates force provider capability and performance requirements into tailored product support to achieve specified and evolving life cycle product support availability, reliability, and affordability parameters. Life cycle sustainment considerations include supply; maintenance; transportation; sustainment engineering; data management; configuration management; human systems integration (HIS); environment, safety (including explosives), and occupational health; protection of critical program information and anti-tamper provisions, supportability, and interoperability. Initially begun during Materiel Solution Analysis (MSA) and matured during the Technology Maturation and Risk Reduction (TMRR) phase, life cycle sustainment planning spans a system's entire life cycle from MSA phase to disposal (DAU, 2017a)

9.0. Financial Management

9.0.1. *Financial management (FM):* The combination of the two core functions of resource management and finance support. (DoD, 2019)

9.0.2. *Resource management (RM):* A financial management function that provides advice and guidance to the commander to develop command resource requirements. (DoD, 2019)

9.0.3. *Finance support:* A financial management function to provide financial advice and recommendations, pay support, disbursing support, establishment of local depository accounts, essential accounting support, and support of the procurement process (DoD, 2019)

10.0. Cost Estimating

10.0.1. *Cost analysis:* An analysis and evaluation of each element of cost in a contractor's proposal to determine reasonableness. (DAU, 2017a)

10.0.2. *Business, cost estimating, and financial management (BCEFM):* Management of acquisition funds including, but not limited to: cost estimating; formulation of input for the Program Objectives Memorandum (POM), the budget, and other programmatic or financial documentation of the Planning, Programming, Budgeting, and Execution (PPBE) process; and budget execution (paying bills). (DAU, 2017a)

10.0.3. *Cost estimate:* An estimate of the cost of an object, commodity, weapon system, or service resulting from an estimating procedure or algorithm. A cost estimate has "context," that is, whether it is the cost to develop and/or procure, and/or to support and/or maintain the item of service and whether it is an incremental, total or Life Cycle Cost, or some other cost perspective. A cost estimate may constitute a single value or a range of values. (DAU, 2017a)

11.0. Auditing

11.0.1. *Audit:* Systematic examination of records and documents to determine adequacy and effectiveness of budgeting, accounting, financial, and related policies and procedures; compliance with applicable statutes, regulations, policies, and prescribed procedures; reliability, accuracy, and completeness of financial and administrative records and reports; and the extent to which funds and other resources are properly protected and effectively used. (DAU Glossary)

11.0.2. *Auditor:* Represents the cognizant audit office designated by the Defense Contract Audit Agency (DCAA) or Service audit activities for conducting audit reviews of the contractor's accounting system policies and procedures for compliance with the criteria. (DAU, 2017a)

12.0. Contract Administration

12.0.1. *Contract administration:* All the activities associated with the performance of a contract from award to closeout. (DAU, 2017a)

12.1. Contracting Actions

12.1.1. *Contracting officer (CO):* A person with authority to enter into, administer, and/or terminate contracts and make related determinations and findings for the U.S. government. In the DoD, these functions are often divided between the Administrative Contracting Officer (ACO) and the Procuring Contracting Officer (PCO). (DAU, 2017a)

12.1.2. Administrative contracting officer (ACO): The government contracting officer (CO) responsible for government contracts administration. (DAU Glossary). Procuring Contracting Officer (PCO): The individual authorized to enter into contracts for supplies and services on behalf of the government by sealed bids or negotiations, and who is responsible for overall procurement under the contract. The term "Procuring" was removed from the Federal Acquisition Regulation (FAR); however, it is still in widespread use to differentiate the buying office Contracting Officer (CO) from the Contract Administrative Office CO, who usually is referred to as the Administrative Contracting Officer (ACO). The FAR uses the term ACO for those actions unique to post contract award; otherwise it uses then generic CO. (DAU, 2017a)

12.1.3. Contracting officer's representative (COR): An individual, including a contractor officer's technical representative (COTR), designated and authorized in writing by the contracting officer to perform specific technical or administrative functions. (DAU, 2017a)

12.2. Contracting Strategy

12.2.1. Contracting strategy: [As part of the Acquisition Strategy] Discuss the planned contract type and how it relates to risk management in each acquisition phase; whether risk management enables the use of fixed-price elements in subsequent contracts; market research; and small business participation. (DoDI 5000.02, 2018)

12.3. Contract Peer Review

12.3.1. Peer reviews: Independent management reviews of supplies and services contracts. Pre-award reviews are conducted on supplies and services contracts; post-award reviews are conducted on services contracts. The Director, Defense Procurement, Acquisition Policy and Strategic Sourcing (DPAP), in the Office of the Under Secretary of Defense for Acquisition, Technology and Logistics (OUSD(AT&L)), conducts peer reviews for contracts with an estimated value of $1 billion or more (including options). DoD components conduct peer reviews for contracts valued at less than $1 billion. (DAU, 2017a)[1]

12.4. Contracting: Acceptance of Deliverables

12.4.1. Expenditure: An actual disbursement of funds in return for goods or services. Frequently used interchangeably with the term outlay. (DAU, 2017a)

13.0. Purchasing

13.0.1. Purchase order (PO): Offer by the Government to buy supplies or services, including construction and research and development, upon specified terms and conditions, using simplified acquisition procedures. (DAU, 2017a; for additional information, see Federal Acquisition Regulation, 2019)

---

[1]   Note that DPAP is now called Defense Pricing and Contracting within the new OUSD(Acquisition and Sustainment).

14.0. Industrial-Base and Supply-Chain Management

14.0.1. *Industrial base (IB):* That part of the total private- and government-owned industrial production and depot-level equipment and maintenance capacity in the United States and its territories and possessions and Canada. It is or shall be made available in an emergency for the manufacture of items required by the U.S. military services [departments] and selected allies. (DAU, 2017a)

14.0.2. *Supply chain management (SCM):* A cross-functional approach to procuring, producing, and delivering products and services to customers. The broad management scope includes subsuppliers, suppliers, internal information, and funds flow (Joint Publication 1-02). SCM provides an intellectual and organizational approach to managing, integrating, and assuring all the elements that affect the flow of materiel to the joint force. Military SCM is the discipline that integrates acquisition, supply, maintenance, and transportation functions with the physical, financial, information, and communications networks in a results-oriented approach to satisfy joint force materiel requirements. (DAU, 2017a; for additional information, see Joint Chiefs of Staff, 2013)

15.0. Infrastructure and Property Management

15.0.1. *Infrastructure:* Generally applicable for all fixed and permanent installations, fabrications, or facilities for the support and control of military forces. (DAU, 2017a)

15.0.2. *Facilities and infrastructure:* One of the 12 Integrated Product Support (IPS) Elements. It encompasses the permanent and semi-permanent real property assets required to support a system, including studies to define types of facilities or facility improvements, location, space needs, environmental and security requirements, and equipment. It also includes facilities for training, equipment storage, maintenance, supply storage, ammunition storage, and so forth. The objective of this IPS Element is to identify, plan, resource, and acquire facilities to enable training, maintenance and storage to maximize the effectiveness of system operation and the logistics support system at the lowest total ownership cost. (DAU, 2017a; for additional information, see DoD, 2016)

16.0. Manpower Planning and Human Systems Integration

16.0.1. *Manpower estimate:* An estimate of the most effective mix of DoD manpower and contract support for an acquisition program. Includes the number of personnel required to operate, maintain, support, and train for the acquisition upon full operational deployment. The estimate is included in the Cost Analysis Requirements Description (CARD) Independent Cost Estimate (ICE) at major milestones. (DAU, 2017a)

16.0.2. *Manpower and personnel (M&P):* The identification and acquisition of personnel (military and civilian) with the skills and grades required to operate, maintain, and support systems over their lifetime. Early identification is essential. If the needed manpower is an additive requirement to existing manpower levels of an organization, a formalized process of identification and justification must be made to higher authority. The terms "manpower" and "personnel" are not interchangeable. Manpower represents the number of personnel or positions required to perform a specific task. Personnel is indicative of human aptitudes (i.e., cognitive, physical, and sensory capabilities), knowledge, skills, abilities, and experience levels that are needed to properly perform job tasks. (DAU, 2017a; for additional information, see DoD, 2016)

17.0. Training and Education

17.1. Training and Education: Government Execution

17.1.1. *Training:* The level of learning required to adequately perform the responsibilities designated to the function and accomplish the mission assigned to the system. (DAU, 2017a)

17.1.2. *Training and training support:* Consists of the policy, processes, procedures, techniques, Training Aids Devices Simulators and Simulations (TADSS), planning and provisioning for the training base including equipment used to train civilian and military personnel to acquire, operate, maintain, and support a system. This includes New Equipment Training (NET), institutional, sustainment training and Displaced Equipment Training (DET) for the individual, crew, unit, collective, and maintenance through initial, formal, informal, on the job training (OJT), and sustainment proficiency training. Significant efforts are focused on NET, which, in conjunction with the overall training strategy shall be validated during system evaluation and test at the individual, crew, and unit level. (DAU, 2017a; for additional information, see DoD, 2016)

17.2. Training and Education: Acquired Systems

17.2.1. *Training and training support* (see 17.1.2, above).

18.0. Disposal

18.0.1. *Disposal:* (1) The second effort of the Operations and Support (O&S) phase as established and defined by DoDI 5000.02. At the end of its useful life, a system shall be demilitarized and disposed of in accordance with all legal and regulatory requirements and policy relating to safety (including explosives safety), security, and the environment. (2) The act of getting rid of excess, surplus, scrap, or salvage property under proper authority. Disposal may be accomplished by, but not limited to, transfer, donation, sale, declaration, abandonment, or destruction. (DAU, 2017a)

Definitions of Acquisition Interface Functions

The following functions involve interfacing with major external activities that affect or influence acquisition. While the external activities are managed outside the domain of acquisition, the acquisition enterprise must manage these interfaces.

19.0. Receive, Inform, and Fulfill Requirements

19.0.1. *Requirement*: 1. The need or demand for personnel, equipment, facilities, other resources, or services, by specified quantities for specific periods of time or at a specified time. 2. For use in budgeting, item requirements should be screened as to individual priority and approved in the light of total available budget resources. (DAU, 2017a)

20.0. Request, Receive, and Respond to Acquisition Intelligence

20.0.1. *Intelligence*: 1. The product resulting from the collection, processing, integration, evaluation, analysis, and interpretation of available information concerning foreign nations, hostile or potentially hostile forces or elements, or areas of actual or potential operations. 2. The activities that result in the product. 3. The organizations engaged in such activities. (DoD, 2019)

21.0. Request and Act Upon Legal Counsel

21.0.1. *Consult legal counsel*: This function occurs on an as-needed basis and reflects times when members of the Program Office or the contracting office consult with legal authorities assigned to the acquisition functions for legal guidance. The lawyers are officially a part of the DOD General Counsel's office, not the Acquisition directorates. (We created this definition based on our prior experience working in the DoD.)

References

4th Estate DACM, "4th Estate Agencies," webpage, undated. As of December 10, 2018:
http://www.doddacm.mil/agencies.html

BLS—*See* U.S. Bureau of Labor Statistics.

Byrne, John P., "Project Management: How Much Is Enough?" *PM Network*, Vol. 13, No. 2, 1999, pp. 49–52.

"CPI Inflation Calculator," webpage, undated. As of December 12, 2018:
https://data.bls.gov/cgi-bin/cpicalc.pl

D'Alessandro, Michael, "Project Management Essentials," *Seattle Daily Journal of Commerce*, November 18, 1999. As of January 19, 2018:
http://www.djc.com/special/design99/10060810.htm

DAU—*See* Defense Acquisition University.

Defense Acquisition University, *DAU Program Managers Tool Kit*, 12th ed., Fort Belvoir, Va., December 2002.

Defense Acquisition University, *Joint Program Management Handbook*, Fort Belvoir, Va.: Defense Acquisition University Press, July 2004. As of December 19, 2017:
http://www.dtic.mil/docs/citations/ADA437767

Defense Acquisition University, *DAU Glossary*, Fort Belvoir, Va., February 9, 2017a. As of March 18, 2018:
https://www.dau.mil/tools/t/DAU-Glossary

Defense Acquisition University, *Defense Acquisition Guidebook*, Fort Belvoir, Va., February 9, 2017b.

Department of Defense Financial Management Regulation 7000.14-R, "Budget Formulation and Presentation (Chapters 4–19)," Volume 2B, Under Secretary of Defense (Comptroller), November 2017. As of December 10, 2018:
https://comptroller.defense.gov/Portals/45/documents/fmr/Volume_02b.pdf

Department of Defense Instruction 7041.04, "Estimating and Comparing the Full Costs of Civilian and Active Duty Military Manpower and Contract Support," July 3, 2013. As of August 3, 2018:
http://www.esd.whs.mil/Portals/54/Documents/DD/issuances/dodi/704104p.pdf

Department of Defense Instruction 5000.02, *Operation of the Defense Acquisition System*, Incorporating Change 4, effective August 31, 2018, January 7, 2015. As of December 17, 2017:
https://www.esd.whs.mil/Portals/54/Documents/DD/issuances/dodi/500002p.pdf?ver=2019-05-01-151755-110

Department of Defense Instruction 8500.01, Cybersecurity, March 14, 2014. As of April 2, 2019:
https://www.esd.whs.mil/Portals/54/Documents/DD/issuances/dodi/850001_2014.pdf

Deputy Chief Management Officer, *2015 Congressional Report on Defense Business Operations*, Washington, D.C.: U.S. Department of Defense, March 15, 2015. As of March 15, 2017:
https://cmo.defense.gov/Portals/47/Documents/Publications/Congressional_Report/2015Congressional_Report.pdf

DoD—*See* U.S. Department of Defense.

Easton, Paul C., "Avoiding Redundant Project-Management Costs in Electronic-Discovery Projects," Legal Project Management, October 4, 2010. As of January 21, 2018:
http://legalprojectmanagement.info/blog/2010/10/avoiding-redundant-project-management-costs-in-electronic-discovery-projects.html

Easton, Paul C., "Avoiding Redundant Project Management Costs in Electronic Discovery Projects," ProjectManagement.com, Newtown Square, Pa: Project Management Institute, August 31, 2011. As of January 21, 2018:
https://www.projectmanagement.com/videos/285154/Avoiding-Redundant-PM-Costs-in-Electronic-Discovery

Federal Acquisition Regulation, Part 2, Subpart 2.1, 2.101—Definitions, November 2019. As of April 2, 2019:
http://farsite.hill.af.mil/reghtml/regs/far2afmcfars/fardfars/far/02.htm#P14_694

HCI—See Human Capital Initiative.

Heywood, George E., and Thomas J. Allen, "Project Controls: How Much Is Enough?" *PM Network*, Vol. 10, No. 11, November 1996, pp. 40–41. As of January 20, 2018:
https://www.pmi.org/learning/library/project-controls-much-enough-4817

Human Capital Initiative, "Defense Acquisition Workforce Key Information: Program Management, as of FY17Q3," U.S. Department of Defense, June 30, 2017. As of December 12, 2018:
http://www.hci.mil/docs/Workforce_Metrics/FY17Q3/FY17Q3_PM_SM.pdf

Ibbs, C. William, and Young-Hoon Kwak, "Assessing Project Management Maturity," University of California, Berkeley, and Florida International University, 1998a. As of January 19, 2018:
http://citeseerx.ist.psu.edu/viewdoc/download?doi=10.1.1.199.7509&rep=rep1&type=pdf

Ibbs, C. William, and Young-Hoon Kwak, "Benchmarking Project Management Organizations," *PM Network*, Vol. 12, No. 2, Newtown Square, Pa: Project Management Institute, 1998b, pp. 49–53.

International Organization for Standardization, "ISO/IEC/IEEE 24765:2010: Systems and Software Engineering—Vocabulary," December 2010. As of April 2, 2019:
https://www.iso.org/standard/50518.html

Joint Chiefs of Staff, Distribution Operations, Joint Publication 4-09, December 19, 2013. As of May 13, 2019:
http://edocs.nps.edu/2012/December/jp4_09.pdf

Jones, Capers, *Applied Software Measurement—Assuring Productivity and Quality*, 2nd ed., New York: McGraw-Hill, 1997.

Kerzner, Harold, *Project Management—A Systems Approach to Planning, Scheduling, and Controlling*, 6th ed., New York, N.Y.: Van Nostrand Reinhold, 1998.

Mahoney, William D., ed., *Means Estimating Handbook*, Kingston, Mass.: R. S. Means Company, 1990.

Marine Corps Systems Command, *MARCORSYSCOM Acquisition Guidebook [MAG]*, updated February 3, 2017. As of December 19, 2017:
https://www.dau.mil/acquipedia/_layouts/15/WopiFrame.aspx?sourcedoc=/acquipedia/PolicyDocuments/the1MAG_Update_03Feb2017.pdf&action=default&DefaultItemOpen=1

Moore, Kevin R., *Using Activity-Based Costing to Improve Performance: A Case Study Report*, Air University, AU/ACSC/125/2000-04, Maxwell AFB: Air Command and Staff College, 2000. As of February 28, 2019:
https://apps.dtic.mil/dtic/tr/fulltext/u2/a393980.pdf

Naval Air Systems Command, *NAVAIR Acquisition Guide 2016/2017*, U.S. Navy, September 22, 2015. As of May 13, 2019:
https://myclass.dau.mil/bbcswebdav/institution/Courses/Deployed/ENG/ENG202/Archives/Student%20CD%20%28April%20FY17%20QTR%203%29/Course%20References/DAG%20and%20Other%20References/NAVAIR%202016_2017%20Acquisition%20Guide.pdf

Nemfakos, Charles, Bernard Rostker, Raymond E. Conley, Stephanie Young, William A. Williams, Jeffrey Engstrom, Barbara Bicksler, Sara Beth Elson, Joseph Jenkins, Lianne Kennedy-Boudali, and Donald Temple, *Workforce Planning in the Intelligence Community: A Retrospective*, Santa Monica, Calif.: RAND Corporation, RR-114-ODNI, 2013. As of September 30, 2018:
https://www.rand.org/pubs/research_reports/RR114.html

Office of the Under Secretary of Defense for Acquisition and Sustainment, "Human Capital Initiatives," website, undated. As of December 10, 2018:
http://www.hci.mil/

OPM—*See* U.S. Office of Personnel Management.

Parker, William, *Defense Acquisition University Program Managers Tool Kit*, 16th ed., Fort Belvoir, Va.: Defense Acquisition University, January 2011. As of December 19, 2017:
http://www.dtic.mil/docs/citations/ADA606320

Secretary of Defense, *Memorandum for the Deputy Secretary of Defense: Establishment of Cross-Functional Teams to Address Improved Mission Effectiveness and Efficiencies in the DoD*, Washington, D.C., February 17, 2017.

Taylor, Michael D., "The Role of the Project Coordinator," *The Silicon Valley Project Management Blog*, Santa Clara, Ca.: University of California at Santa Cruz Extension in Silicon Valley, May 20, 2008. As of March 5, 2019:
http://svprojectmanagement.com/the-role-of-the-project-coordinator

Under Secretary of Defense (Comptroller), "DoD Budget Request," webpage, undated. As of December 10, 2018:
https://comptroller.defense.gov/Budget-Materials/

Under Secretary of Defense for Acquisition, Technology, and Logistics, *Performance of the Defense Acquisition System: 2013 Annual Report*, Washington, D.C.: U.S. Department of Defense, 2013. As of March 9, 2018:
http://www.dtic.mil/docs/citations/ADA587235

Under Secretary of Defense for Acquisition, Technology, and Logistics, *Performance of the Defense Acquisition System: 2014 Annual Report*, Washington, D.C.: U.S. Department of Defense, 2014. As of March 9, 2018:
http://www.dtic.mil/docs/citations/ADA603782

Under Secretary of Defense for Acquisition, Technology, and Logistics, *Performance of the Defense Acquisition System: 2015 Annual Report*, Washington, D.C.: U.S. Department of Defense, 2015. As of March 9, 2018:
http://www.dtic.mil/docs/citations/ADA621941

Under Secretary of Defense for Acquisition, Technology, and Logistics, *Performance of the Defense Acquisition System: 2016 Annual Report*, Washington, D.C.: U.S. Department of Defense, 2016a. As of March 9, 2018:
http://www.dtic.mil/docs/citations/AD1019605

Under Secretary of Defense for Acquisition, Technology, and Logistics, *Department of Defense Acquisition Workforce Strategic Plan, FY 2016–FY 2021*, December 2016b. As of March 18, 2018:
http://www.hci.mil/docs/Policy/Legal%20Authorities/DoD_Acq_Workforce_Strat_Plan_FY16_FY21.pdf

USASpending.gov, website, undated. As of December 12, 2018:
https://www.usaspending.gov/#/

U.S. Bureau of Labor Statistics, Division of Occupational Employment Statistics, "May 2016 National Occupational Employment and Wage Estimates—United States," Washington, D.C., May 2017. As of February 28, 2018:
https://www.bls.gov/oes/current/oes_nat.htm

USD(AT&L)—*See* Under Secretary of Defense for Acquisition, Technology, and Logistics.

U.S. Department of Defense, *Military Handbook: Configuration Management Guidance*, MIL-HDBK-61A(SE), February 7, 2001. As of April 2, 2019:
http://acqnotes.com/Attachments/MIL-HDBK-61A%20(SE)Configuration%20Management%20Guidance.pdf

U.S. Department of Defense, *Technology Readiness Assessment (TRA) Deskbook*, Washington, D.C.: Director, Research Directorate (DRD), Office of the Director, Defense Research and Engineering (DDR&E), July 2009. As of April 2, 2019:
https://www.skatelescope.org/public/2011-11-18_WBS-SOW_Development_Reference_Documents/DoD_TRA_July_2009_Read_Version.pdf

U.S. Department of Defense, *Department of Defense Interface Standard: Electromagnetic Environment Effects Requirements for Systems*, MIL-STD-464C, December 1, 2010. As of April 2, 2019:
https://snebulos.mit.edu/projects/reference/MIL-STD/MIL-STD-464C.pdf

U.S. Department of Defense, *Product Support Manager Guidebook*, Washington, D.C.: Assistant Secretary of Defense for Logistics and Materiel Readiness, April 2016. As of April 2, 2019:
https://www.dau.mil/guidebooks/Shared%20Documents%20HTML/PSM%20Guidebook.aspx

U.S. Department of Defense, *Department of Defense Risk, Issue, and Opportunity Management Guide for Defense Acquisition Programs*, Washington, D.C.: Office of the Deputy Assistant Secretary of Defense for Systems Engineering, January 2017. As of April 2, 2019:
https://www.acq.osd.mil/se/docs/2017-rio.pdf

U.S. Department of Defense, *DoD Dictionary of Military and Associated Terms*, February 2019. As of April 2, 2019:
https://www.jcs.mil/Portals/36/Documents/Doctrine/pubs/dictionary.pdf

U.S. Office of Personnel Management, "Classification and Qualifications: Classifying General Schedule Positions," webpages, Washington D.C., undated(a). As of February 22, 2018:
https://www.opm.gov/policy-data-oversight/classification-qualifications/classifying-general-schedule-positions/

U.S. Office of Personnel Management, "FedScope," webpage, undated(b). As of December 10, 2018:
https://www.fedscope.opm.gov/

U.S. Office of Personnel Management, *Handbook of Occupational Groups and Families*, Washington D.C., December 2018. As of May 15, 2019:
https://www.opm.gov/policy-data-oversight/classification-qualifications/classifying-general-schedule-positions/occupationalhandbook.pdf